Technical Writing
THEORY AND PRACTICE

Edited by

Bertie E. Fearing and
W. Keats Sparrow

WITHDRAWN

LIBRARY OF MOUNT ST. MARY'S

THE MODERN LANGUAGE ASSOCIATION OF AMERICA
NEW YORK   1989

Copyright © 1989 by The Modern Language Association of America

Library of Congress Cataloging-in-Publication Data

Technical writing : theory and practice / edited by Bertie E. Fearing
    and W. Keats Sparrow
        p.   cm.
    Bibliography: p.
    Includes index.
    ISBN 0-87352-180-3 ISBN 0-87352-181-1 (pbk.)
    1. Technical writing.   I. Fearing, Bertie E.   II. Sparrow, W.
Keats (Wendall Keats), 1942- .
    T11.T337   1989
    808′.0666—dc20   89-12477

Second printing, 1990.

Published by the Modern Language Association of America
10 Astor Place, New York, NY 10003-6981

Contents

Preface

THE NATURE AND SCOPE of this anthology emerged from a wide-ranging survey of instructors. The book thus reflects the thinking of many technical writing specialists throughout the country and beyond, among them the members of the Council for Programs in Technical and Scientific Communication and the leaders of the Association of Teachers of Technical Writing. (As used in this anthology, *technical writing*, or its synonym *technical communication*, is a general term referring to the kinds of practical discourse found in industrial and technological settings—manuals, proposals, documentation, mechanism descriptions, and the like—as well as to such related communication skills as oral presentations, graphics, and layout and design. An ill-fitting and confusing term as conventionally used, it often overlaps with and encompasses the writing of such fields as business and science—letters, memorandums, laboratory reports, scientific articles, and the like.)

Most of those who advised us wanted the first technical writing volume published by the Modern Language Association to be broadly based, addressing a variety of theoretical and practical issues confronting instructors who teach advanced as well as introductory courses, plan curricula, and direct programs. In other words, the anthology should not have a narrow focus and deal solely with a single issue or theme; it should not be so theoretical as to be impractical; and it should not be directed to novice teachers. Instead, it should represent a wide range of views from teachers and practitioners of technical writing. Its essays should be new ones written especially for the volume and based on current research.

We asked our advisers to indicate topics they would like to know more about and to suggest the best authors for those topics. After tabulating the 123 responses, we identified the most representative topics and invited the recommended authors to submit essays. We made our selections from the essays submitted and then commissioned a few additional ones to fill obvious gaps or special needs.

The result is an essay collection whose four parts deal with issues that many members of the profession currently deem important: (1) On the History and Theory of Technical Writing, (2) On the Composing Process in Corporate Settings: From Industry to Academe, (3) On the Process and Product of Technical Writing: Contemporary Perspectives, and (4) On Teaching Technical Writing: Current and Recurrent Issues.

A careful reading of the thirteen essays in these four divisions will disclose that they do, in fact, follow our advisers' suggestions. They are varied: Some look to the past (Souther), some to the future (Tebeaux), others to both (Allen); some apply theory (Miller) and current research (Redish and Schell, Jordan, Lay) to practice, while others examine several current issues that advanced teachers will find provocative or useful, including collaborative writing (Grice, Debs, Selzer), textbook resources for advanced courses (Warren), and instructional approaches (Mitchell and Smith). The essays reveal much of the current thinking of established members of the profession and of promising newer members.

The volume thus provides what leaders in the field believe the profession needs in the latest of its growing number of anthologies—an examination of various issues, both theoretical and practical, facing teachers.

Moreover, and perhaps more important, collectively the essays mirror the status of the teaching of technical writing as the twentieth century draws to a close. Perhaps the most arresting image they reflect is that of a profession with many divergent paths. For example, should we retain our practical orientation or accept our theoretical drift (Souther)? If we opt for practicality, do we adhere to the "low" or "high" model (Miller)? Should we teach the writing process as a solitary activity or as the collaborative effort found in industry (Grice, Debs, Selzer)? Should we reconsider audience analysis in the light of protocol analysis and usability testing (Allen, Redish and Schell), and should we put aside symmetry for asymmetry as our premise for page layout (Lay)? Are we to continue promoting impressionistic precepts about style, or are we to use empirical findings (Jordan, Thompson)? And is the better pedagogical model prescriptive (Mitchell) or heuristic (Smith)?

The sheer number of problematical choices noted in the essays—far more than those mentioned in this preface—suggests that their resolution will lead imminently to revisions in our professional assumptions. The teaching of technical writing thus appears to be on the verge of a transformation whose outcomes will soon be found in our courses, curricula, and textbooks.

While the essays herald uncertainties and upheaval, they nevertheless give an encouraging view of the profession as we approach our centenary. With our own journals and organizations, we have emerged as a recognized discipline and have finally come of age in English studies. We are engaging lustily in many kinds of research—historical, theoretical, survey, empirical. Perhaps as a consequence, we also appear to be moving toward a consensus about our theoretical foundations and about the proper relation between our teaching and industrial practices.

Just as important, we are reexamining our premises, refining some and replacing others as we do so. Faster than teachers in any other area of English studies, we are becoming interdisciplinary and even intercultural. We are also recognizing our need to adapt to the new technologies and, at the same time, to guide the discourse of those technologies.

Clearly, as we look toward the twenty-first century, we are a heady profession undergoing a rigorous and healthy self-assessment and are intent on development, even though we have just arrived. We have no love for inertia and the status quo.

In closing, we wish to thank Virginia Book and other members of the Council of Programs in Technical and Scientific Communication and Donald H. Cunningham and other members of the Association of Teachers of Technical Writing for their guidance in shaping the contents of the book. We especially thank Merrill Whitburn of Rensselaer Polytechnic Institute for his valuable suggestions; Dwight Stevenson of the University of Michigan for his thoughtful advice; James Paradis of Massachusetts Institute of Technology for his encouragement; Kevin Rouse and Margaret Wirth of East Carolina University for their bibliographic, statistical, and editorial assistance; Sally Funke Lawrence of East Carolina University for compiling the index; John S. Patterson of East Carolina University for his help with matters too numerous to mention; and Joe Gibaldi of the Modern Language Association for inviting technical writing to become a part of MLA's publication program.

BEF
WKS

W HILE THE TEACHING OF technical writing is a relative newcomer to English studies, it is not a discipline without a heritage. As much recent research has disclosed, the discipline has a rich and revealing past. Rich as that background is, however, it still has not provided us with a full understanding of what is meant by the "practicality" of technical writing—that is, whether it should be "low" in the how-to sense or "high" in the Aristotelian sense. The essays in this section address these two possibilities.

Written by one of the most venerable members of the profession, the first essay is an insider's history of the teaching of technical writing. James W. Souther reviews the achievements of leading technical writing teachers—from those who established new and distinct service courses in the 1920s and 1930s to those who developed degree programs toward the end of the century. The essay helps establish a sense of tradition and concludes with the assessment that, on the eve of the twenty-first century, we are a mature and expanding field but one facing numerous challenges. Later essays in this volume explore a number of those challenges more fully.

In the second essay, Carolyn R. Miller differentiates between the two senses of practicality that have posed an enduring dilemma for our field—the pejorative sense of low forms of practice versus the laudatory sense of Aristotle's praxis, an intellectual activity intended to awaken and enhance critical consciousness. This dichotomy—echoed in Jack Selzer's study of corporate composing processes in part 2 and in the pedagogical debate by John H. Mitchell and Marion K. Smith in part 4—is responsible for what Miller terms the "uneasy relationship between nonacademic practice and academic instruction [that] has been part of academic discussions about technical writing from their beginnings. . . ." In positing a theoretical base for the profession, Miller favors the latter sense— "a locus for questioning, for criticism, for distinguishing good practice from bad."

The two general essays in this part do not aim at definitiveness. Rather, they attempt to examine our history and theory in such a way as to improve understanding, provoke further thought, and elicit additional research and analysis. They also establish a context for the essays in parts 2, 3, and 4 of the book.

Teaching Technical Writing: A Retrospective Appraisal

JAMES W. SOUTHER

TECHNICAL COMMUNICATION EMERGED in the 1980s as a discipline possessing both professional maturity and academic sophistication. Although many individuals and events have contributed to the present stature of the field, three developments have particular significance: the growth in technical communication academic programs, the expanding roles within the technical communication profession itself, and the quality of technical communication research.

1. The stature and credibility of the field have been enriched by the growth of technical communication academic offerings—measured by the number of classes, faculty members teaching these courses, degree programs, and English departments' involvement. For example, in 1976 the Society for Technical Communication listed only 19 academic degree programs in technical communication (Pearsall and Sullivan). In 1981 this number grew to 28 (Pearsall, Sullivan, and McDowell), and by 1985 the number had swelled to 58 (Kelley et al.). As these figures show, "the quantity of academic programs in technical communication almost doubled from 1976 to 1981 and then doubled again from 1981 to 1985" (Kelley et al. x). The growth is also reflected in the activity of our professional societies. In 1973 the Association of Teachers of Technical Writing was little more than a dream of John Harris, John Mitchell, David Carson, Don Cunningham, and others. Today it is a thriving organization and the publisher of the *Technical Writing Teacher*, one of the two primary journals in the field. A few years ago technical communication papers were not found on the programs of our major societies; today they are included in sizable numbers at meetings of the NCTE, the CCCC, and the MLA.

2. The second factor contributing to this stature and credibility has been the rapid expansion of the technical communication field itself. The number of professionals has expanded almost in direct relation to the spread of computer technology. For example, the Society for Technical Communication has grown from 7,300 members in 1984 to over 12,000 in 1988. Moreover, the scope of the profession has been broadened by three mandates requiring effective communication of technical information to the public: (a) the environmental legislation, both federal and state, requiring that understandable environmental-impact statements be made available to the public, (b) the consumer movement

that increased corporate and governmental responsibility for providing reliable information in plain language, and (c) the advent of the personal computer, which places the usability of software documentation at the center of marketing strategy.

3. But as important as academic growth and professional expansion are to the stature of technical communication, the primary measure of the growing maturity of the field is the quality of its recent research, for example, the work conducted by American Institutes for Research Document Design Center, under Janice Redish, and the related research activity at Carnegie Mellon University. The center's application of research to solve communication problems in government and industry and the direction of the research at Carnegie Mellon have provided our field with significant new dimensions. The quality of present scholarship, however, is most clearly revealed in such recent anthologies as *New Essays in Technical and Scientific Communication: Research, Theory and Practice*, edited by Paul Anderson, John Brockmann, and Carolyn Miller; *Writing in Nonacademic Settings*, edited by Lee Odell and Dixie Goswami; *Research in Technical Communication: A Bibliographic Sourcebook*, edited by Michael Moran and Debra Journet; and this, the latest anthology, edited by Bertie Fearing and Keats Sparrow.

Yet, with all the increased activity, are we teaching better classes today? Are we providing our students with a more enriched education in technical writing than they would have received in the past? Are today's classes better than those offered by Henrietta Tichy or Earl Britton some twenty years ago? Has teaching kept up with the sophistication of our scholarship, or has our scholarship had more impact on faculty members than on students? Above all else, effective teaching requires an awareness of the practices and requirements of the profession, yet much of our research tends to be isolated from the profession. Perhaps the ingredient necessary to more effective teaching is greater awareness of the technical communication profession itself. Technical communication is, after all, a professional program. Application and practice are just as essential as academic theory.

Despite the increased stature we have experienced, we have somehow failed to develop a sense of tradition, a knowledge of what has come before and how it relates to us. As much as we may individually value historical continuity, most of us know little of past developments in technical communication. Since movements in our field today are, in large measure, a reflection of yesterday, this essay focuses on some of the milestones from the early 1900s to the present.

The essay concentrates on five approaches to the teaching of technical writing developed during this period: language, rhetoric, product, process, and hybrid approaches. The first approach sees effective use of language, grammar, and style as the focus for the technical writing course; the second centers on rhetorical principles and modes of discourse; the third views technical writing as a series of different forms and products; the fourth teaches technical writing as a process that produces writing for different audiences; and the fifth combines these approaches into an integrated pattern.

1920s to 1950

When many of us came to technical writing after World War ii, considerable literature was available, including a number of textbooks. In 1908, T. A. Rickard wrote what was likely the first book on technical writing, *A Guide to Technical Writing*. Rickard's long experience as an editor of mining journals gave his book a language emphasis, focusing on usage and style. Although this work was intended for practicing engineers, his next book, *Technical Writing* (1920), was a textbook that retained the language emphasis. In between the publishing dates of Rickard's two books, Samuel Earle of Tufts College published what Robert J. Connors calls "the first genuine technical writing textbook written for use in college courses." *The Theory and Practice of Technical Writing* (1911) was by "the man who, more than any other, deserves the title of Father of Technical Writing" (332). By classifying technical writing into narration, description, exposition, and instruction, Earle provides the first rhetorical, or modes-of-discourse, book.

Sada Harbarger's *English for Engineers* (1923) was the first textbook to present the product approach, by focusing on the technical forms of reports and letters—an approach that remains the basis for many textbooks today. Harbarger appears to be following a pattern that had been established earlier in books on business writing. In 1924, Ronald Press published Ralph Fitting's *Report Writing*, which strengthened the products approach by emphasizing both periodic and examination reports and also included preliminary, investigation, fieldwork, and logical-formulation reports. And, in 1925, Sam Trelease and Emma Yule published *Preparation of Scientific and Technical Papers*, which guided writers through the publication process.

The 1920s, then, witnessed the development of both the language and the product approaches as well as the beginning of the rhetorical approach. Although the language approach developed first, the product approach gained strength and predominated by the early 1930s. During the 1930s and early 1940s, three useful and influential books appeared that those of us entering the field in the years immediately following World War ii were to find most helpful.

The first was Carl Gaum and Harold Graves's *Report Writing*, published in 1929, followed by a second edition in 1942 and a third with Lynne Hoffman as coauthor in 1950. This popular work, which was used for about twenty-five years, was a practice-oriented work emphasizing form and structure, with especially rich examples. Although it contained chapters called "Fundamentals of Composition" and "Requirements of Style," its primary focus was on products.

The second, *The Engineer's Manual of English*, written by W. O. Sypherd and Sharon Brown, appeared in 1933 and in a second edition in 1943 with Alvin Fountain as a third author. As Fountain points out—and as this book illustrates—"a technical-forms approach of a rigid and mechanical sort had become all but absolute by the late thirties" (Connors 338). In 1957, a revised version of the book appeared under the new title *Manual of Technical Writing* and with yet a fourth coauthor, U. E. Gibbons.

The third of the highly influential books was J. Raleigh Nelson's *Writing the Technical Report*, published in 1940, with a second edition in 1947. Nelson drew on his experience, starting in 1923, as editor of publications in the department of engineering research at the University of Michigan. Nelson says of his own experience:

> My most important function was not inserting commas or revamping awkward sentences but helping the weary investigator plan the presentation of his results to meet the requirements of the man paying the bill. . . . To those years of planning how best to shape and organize research material for actual use, I owe the conviction that became the central thesis of my course and is the most inescapable emphasis of this book, that a report must be designed for a specific purpose and that its success must depend largely on how well it is adapted for all the uses to which it is likely to be put. (x)

Nelson's book is in many ways as valuable today as it was when first published. His experience, his insights, and his ability to translate theory into professional practice are as sound today as they were when he set them down some fifty years ago. These days, we make much of J. C. Mathes and Dwight Stevenson's insistence on the design of reports and on invention. As important as their contribution is, however, it is clearly related to the design principles Nelson espoused at Michigan in 1940. His was the first book to suggest the process approach, with its emphasis on design.

The period from the 1920s to the 1950s, then, saw the early development of four of the approaches. Although the language approach was developed first, the product approach soon became dominant. The early efforts of Earle to establish the rhetorical approach and those of Nelson to establish the process approach were not widely adopted, yet they became the ground on which future efforts to establish these approaches would stand.

The 1950s

The 1950s brought new books, but the decade will be remembered for the number of new teachers who came to technical writing. Although there are many more technical writing teachers today, the percentage of those teaching technical writing rose significantly in those years. Moreover, nearly all those who had been the primary contributors before World War II were no longer active in the field by the late 1950s and early 1960s. A second generation had arrived.

In the 1950s we faced conditions greatly different from those that we know today. Even though faculty members were under the same pressures to publish as they are now, there were no journals in our field. Technical writing articles were not yet accepted in traditional English publications, and Jay Gould's *Journal of Technical Writing and Communication* and ATTW's *Technical Writing Teacher* were not yet in existence. The Society for Technical Communication, merging with several other organizations and working its way through the several name changes brought about by those mergers, was not a real force until the second half of the 1950s.

Most technical writing teachers were in English departments in technical institutes or engineering colleges (MIT, Rensselaer Polytechnic Inst., Georgia Inst. of Tech, Newark Coll. of Engineering, Colorado State Univ.) or in special departments within colleges of engineering in full-service universities (Univ. of Michigan and Univ. of Washington are the only two extant). Although we maintained memberships in the traditional organizations such as MLA, NCTE, and CCCC, we found little interest in technical writing there. To find professional support, we had to turn to other societies. Because of our involvement in engineering education, the English Division of the American Society for Engineering Education (ASEE) became our primary source of professional interaction and support. In June of each year, fifteen or twenty technical writing professors—among them Alvin Fountain of North Carolina State University, George Crouch of the University of Pittsburgh, Sterling Olmstead then of Rensselaer Polytechnic Institute, Jim Pitman and Herman Estrin of Newark College of Engineering, and W. C. Brown of General Motors Institute—would gather at the annual conference to share experiences and problems.

By the mid 1950s, the Society for Technical Communication, then known as the Society of Technical Writers and Editors, was becoming an effective, influential organization, and its annual International Technical Communication Conference (ITCC) presented an opportunity for those of us in higher education to meet with technical communication practitioners and to establish a dialogue that would enrich our teaching of technical writing for years to come. Here, Earl Britton, John Walter, Bob Sencer, Henrietta Tichy, Richard Davis, John Mitchell, Frank Smith, Jay Gould, Ben Weil, Myron White, and others shared ideas with technical communication professionals. The STC journal *Technical Communication* was the first devoted to our field. In 1958, the Institute of Electrical and Electronic Engineers (IEEE) established its *Transactions on Engineering Writing and Speech*, now called *Transactions on Professional Communication*, the second important journal for those of us in technical writing.

Five books that appeared in the late 1940s had a particularly strong impact during the early 1950s. Four public lectures by Reginald O. Kapp, dean of the Faculty of Engineering at the University of London, led to an excellent book entitled *The Presentation of Technical Information*. This book, with a few additions by a later editor, is still available and, in spite of its age, contains some interesting insights. *Writing Scientific Papers and Reports*, by Walter Paul Jones of Iowa, was a popular text for many years. *Report Preparation* (1948; 2nd ed. 1951), by Frank Kerekes and Robley Winfry of Iowa State University, was an excellent book rich in real examples from the world of work. William George Crouch and Robert L. Zetler's *A Guide to Technical Writing* emphasized both written and oral reporting in the professional setting. *Engineering Reports*, by Lisle Rose (who was the editor of the ASEE journal) and coauthors Burney Bennett and Elmer Heater, was the last of the books growing out of the 1940s. These five books, product-oriented except for Kapp's lectures, provided the foundation for the developments of the 1950s.

The 1950s witnessed the emergence of books by the new generation of technical communication teachers who came to the field at the close of World War II. The first of these appeared in 1952, Joseph Ulman's *Technical Reporting*, and with Jay Gould as coauthor for later editions, this work had a long and useful life. Two years later, *Technical Writing*, by Gordon Mills and John Walter, appeared and quickly became the standard work, still used widely today. Close behind was *Modern Technical Writing*, by Theodore Sherman of the University of Idaho. This work is still available in a 1983 edition with Simon Johnson of Oregon State University as coauthor. The year 1956 saw the introduction of *Writing Useful Reports* by W. C. Brown and Robert Tuttle of General Motors Institute. In 1957, my own *Technical Report Writing* appeared; heavily influenced by the work of Porter Perrin, then at the University of Washington, it was the first of the process-organized books (a second edition appeared in 1977 with Myron White as coauthor). The following year saw the publication of Margaret Blickle and Kenneth Houp's *Reports for Science and Industry* and a unique book entitled *Technical Editing* by Benjamin Weil et al. This remarkable group of works represents early efforts to combine the various approaches, especially language, rhetoric, and product. Five of these books are still available today, in later editions of course. Articles of the period are more difficult to locate because they appeared in such a wide variety of engineering, professional, and trade journals. They can be found in the pages of *Chemical Engineering, Mechanical Engineering, Product Engineering, Machine Design*, the *Journal of Engineering Education*, and similar publications.

Technical communication thus began to grow as a discipline in the 1950s. Approaches to teaching technical writing were developed more fully. No longer one distinct approach, the language approach was incorporated into all books. The product approach remained dominant, although the rhetorical approach was becoming stronger. The process approach, still in its infancy, had a strong appeal in industry. For example, General Electric centered its in-house course on the process approach; IBM initiated a technical-staff-audience study as a result of the process emphasis on audience analysis (Souther, "Identifying the Informational Needs of Readers"); and Westinghouse conducted a study entitled "What Management Wants in the Technical Report" (Souther), the findings of which received particularly wide use, for the company made some ten thousand reprints available to technical writing teachers for use in their classes. The study has recently been made available to today's teachers in a special issue of *IEEE Transactions on Professional Communication* edited by James Hill.

The 1960s

The 1960s saw a continuing flow of new books on technical writing. More people entered the field, and a few English departments became active in the field. Although engineering enrollment declined, technical writing somehow escaped the most serious consequences. The drop in traditional enrollment in the service courses, in fact, may well have intensified the efforts to develop degree curricula, which began to appear in the 1960s. The NCTE and the CCCC started to include sessions on technical writing, primarily because of the efforts of Herman Estrin and John Harris. The STC would soon join with John Wiley and Sons in publishing the Wiley Human Communication Series under the editorship of Kenneth Tong. Although the task of putting together a two-volume work in that series was under way during the late 1960s, the *Handbook of Technical Writing Practices*, edited by Stello Jordan, would not appear until 1971. It is a monumental and valuable work, but unfortunately it needs revising today.

Because such a large number of books appeared during the 1960s, only those that had the greatest impact or suggested new approaches will be discussed here. The first of these, *The Craft of Technical Writing* by Daniel Marder of the University of Pittsburgh, appeared in 1960. One of the better rhetorical-approach books, it had significant impact on the development of the ESL and technical writing materials by Louis Trimble, Larry Seleker, and Thomas Huckin, who were then at the University of Washington. Herman Weisman's *Basic Technical Writing* has just appeared in its fifth edition, and his *Technical Correspondence* is still the best book of its type. John Mitchell's *Handbook of Technical Communication* was also published in 1962, and both his and Weisman's books were fine examples of the rhetorical approach blended with the language approach. In 1962, Robert Rathbone of Massachusetts Institute of Technology and James B. Stone coauthored *A Writer's Guide for Engineers and Scientists*. This interesting and useful book, enriched by Rathbone's broad consulting practice, emphasizes rhetoric, style, and the process. Another interesting book, with the title *How to Write Better and Faster*, by Terry Smith, appeared in 1965. Smith's book reflects his experience at Westinghouse by emphasizing planning, writing, and editing along with illustrations and production—thus blending the process and product approaches.

The four remaining books of the ten selected for comment in some ways made even more significant contributions. The first is *Effective Writing for Engineers, Managers, Scientists*, by Henrietta Tichy of Hunter College, one of the outstanding books in our field and every bit as good today as it was then because of Tichy's wide consulting experience and talent for cutting to the center of an issue. The second book is *Analytical Writing*, by Thomas Johnson, unfortunately now out of print. Johnson's article "How Well Do You Communicate?" offers a short version of an important part of the book; and a recent article coauthored by Mary Coney and me, "Analytical Writing Revisited," keeps some of this book alive.

In giving us *Reporting Technical Information*, Kenneth Houp and Thomas Pearsall set the stage for the 1970s. This book had the greatest impact on the teaching of technical writing, although Mills and Walter's *Technical Writing* runs a strong second. Houp and Pearsall's work employs several approaches: Part 1 combines the process and the rhetorical approaches; part 2 stresses the products approach, including the oral report; and part 3 is a short handbook that reflects the language approach. This book is the first of the excellent hybrids, those books that so effectively combine the various approaches, and it does so with solid content and presentation. William Gallagher's *Report Writing for Management*, an excellent book full of practical, workplace suggestions, contains many valuable insights from the author's years of experience as a communications consultant. It is another important process book useful for its insight into practices and problems in industry.

If technical writing began to grow up in the 1950s, it emerged from the 1960s with all the self-confidence of young adulthood. It established audience analysis as a basis for writing. This concept evolved from the process approach and the research on the reading habits and informational needs of management conducted by Westinghouse; it was given even greater emphasis by the publication of Thomas Pearsall's *Audience Analysis for Technical Writing*. The need for situational analysis was well established by the end of the 1960s.

One of the most dramatic developments of the 1960s was the growth of the short course or institute, a two- or three-day concentrated course that made the expertise of technical writing faculty members available to industrial and governmental professional employees. Rensselaer Polytechnic Institute, the University of Michigan, Massachusetts Institute of Technology, Colorado State University, and the University of Washington have the longest traditions in offering such courses. Some programs developed menus of courses from which an organization could select; the faculty would then adapt the course to the organization's specific set of problems. First taught on campus, such courses later moved into the industrial or governmental setting. For an organization with many potential students, it was more economical to bring the teacher to the organization than to transport all the students to a campus. The resulting involvement with nonacademic organizations was also an important learning experience for those of us teaching technical communication, an experience that allowed us to look at industrial and governmental writing problems on a broad organizational scale. In a real sense, the 1960s was the decade in which the writing short course became a way of life.

In many ways, however, the growth of degree programs in technical communication was the most significant development of the period. Before the 1960s, technical writing courses were service courses, courses offered to students majoring in other fields, such as engineering, agriculture, business, or science. Although many faculty members still teach service courses today, some also teach in degree programs and have the rare experience of teaching students who are majoring in the field.

The first degree program was the master's program started at Rensselaer Polytechnic Institute in 1958, and this program, then under Jay Gould and now under David Carson, is still one of the best. Rensselaer soon developed a doctoral option whose graduates, as former chair Merrill Whitburn is pleased to point out, are top professionals and faculty members across the country. In the 1960s, both Massachusetts Institute of Technology and the University of Michigan offered graduate programs. Colorado State University was one of the first to offer a degree in technical journalism. Erwin Steinberg, at Carnegie Institute's Margaret Morrison College, established a small undergraduate program. Although a number of other institutions were investigating or planning programs during the 1960s, they would number only ten in the early 1970s.

The 1950s clearly defined the nature and functions of the technical writing service course, but the 1960s saw technical writing draw more closely to professional practice in industry and government. The interactive dialogue is reflected in the creation of degree programs, in the shift of our important society affiliations, in the development of short courses, and in the growth of consulting as a faculty activity.

Of these activities, only the shift in society affiliation needs additional comment. Through the 1950s, the English division of the ASEE had been the primary association. By the late 1950s, the STC and the IEEE Professional Communication group offered new and important professional alignments. In the 1960s, the NCTE and the CCCC began to include technical writing sessions in their programs. By the early 1960s, ASEE participation was waning, for by then technical communication faculty members had begun to turn to other, more meaningful associations: the STC, the IEEE, the NCTE, and the CCCC.

1970s to the Present

The 1970s and the 1980s have seen the fruition of much of what started in the 1960s, for these two decades brought more students, more classes, more teachers, more English department participation, more degree programs, more associations and journals, more professional involvement, more research, and more books. The 1970s and the 1980s are, of course, closer and better known to most of those now active in the field; consequently, the coverage here can concentrate on a few major developments.

A most significant development took place in 1970: from the start, Jay Gould's *Journal of Technical Writing and Communication* was the most respected in the field. In 1973, the Association of Teachers of Technical Writing (ATTW) was formed as an MLA allied organization, and its journal, the *Technical Writing Teacher*, began publication. Although both the association and its journal struggled for a few years, both have gained substantial stature. When these two journals are added to those of the STC and the IEEE, the 1970s provided four professional journals. Today two other journals, the *Journal of Advanced Composition* and the Iowa State University *Journal of Business and Technical Communication*, add to our scholarly base. Because of the growing importance of computer interaction, we must also turn to a number of publications new to us: for example, *Human Factors*, a publication of the Human Factors Society; the *SIGCHI Bulletin*, a publication of the Special Interest Group in Computer Human Interaction of the Association for Computing Machinery; and the *Journal of Applied Psychology*.

A second development in the 1970s was the emergence of resource books for teachers of technical writing, beginning in 1975 with Thomas Pearsall's *Teaching Technical Writing: Methods for College English Teachers* and Donald Cunningham and Herman Estrin's *The Teaching of Technical Writing*. These first works were shortly followed by Thomas Sawyer's *Technical and Professional Communication: Teaching in the Two-Year College, Four-Year College, Professional School* in 1977, Dwight Stevenson's *Courses, Components, and Exercises in Technical Communication* in 1981, and W. Keats Sparrow and Nell Ann Pickett's *Technical and Business Communication in Two-Year Programs* in 1983.

Many new degree programs were established during the 1970s: University of Minnesota, St. Paul, University of Washington, New Mexico State University, Miami University of Ohio, Michigan Tech University, Bowling Green State University, Oklahoma State University, Carnegie Mellon University, Drexel University—to name a few. By the mid 1970s, Tom Pearsall and Tom Warren created the Council of Programs in Technical and Scientific Communication (CPTSC). Other programs appeared in the late 1970s, and the pace has accelerated, from a half dozen programs in the 1960s to almost twenty in the 1970s to nearly sixty today.

Research also grew, not only in volume but, more important, in sophistication during the 1970s and 1980s, and every sign points to expansion in the 1990s. A number of important scholars have contributed, and while all the work cannot be mentioned here, two or three major thrusts need comment: (1) the consulting of the Document Design Center, (2) the investigations of Carnegie Mellon University, and (3) the efforts of individuals like Michael Halloran, Carol Lipson, Victoria Mikelonis, Carolyn Miller, Lee Odell, Jim Paradis, Judy Ramey, Marilyn Samuels, Jack Selzer, Jan Spyridakis, and others. These efforts contribute to our current knowledge and provide a basis on which our future research can build. Our task is to keep abreast and to become participants. But our research must not become isolated from the profession, for technical communication, above all else, is a professional field.

Certainly the 1970s and the 1980s saw an increasing number of new books, too many to mention here. The principal characteristic of these more recent books is that they combine various early approaches. Most books published in the 1970s are examples of the hybrid approach established by Houp and Pearsall in the 1960s. In many, the product approach is maintained, the rhetorical approach is expanded, and the language approach is included. Many of our recent authors—Deborah Andrews and Margaret Blickle, Ron Blicq, Marya

Holcombe and Judith Stein, Charles Stratton, and Tom Warren, for example —have produced excellent "hybrid" books. But of the recent works, John Lannon's *Technical Writing* remains one of the leading books in our field. It and Michael Keene's *Effective Professional Writing* are perhaps the best recent examples of the hybrid approach. The logical and functional balance of the various approaches provides abundant material from which each teacher can choose that which he or she wants to emphasize. Both have process, both have rhetoric, both have product, and both have language.

Three other books deserve special attention for continuing the process approach and building on the earlier works (e.g., Hays; Gallagher). The first is J. C. Mathes and Dwight Stevenson's *Designing Technical Reports*; the second, Nancy Roundy and David Mair's *Strategies for Technical Communication*; and the third, Paul Anderson's *Technical Writing*. The first was certainly one of the most influential books of the 1970s; more than any other book, it brought the process approach to new teachers of technical writing. The second helps the teacher apply the process to different forms of writing. The third, with its focus on reader impact, helps us understand the nature of technical communication. Each carries situational analysis into the decision making that is technical writing.

Today's authors blend the best from each of these approaches into a useful and functional approach that serves teacher and student alike. Yet each is indebted to those who came before.

Today, technical communication has achieved stature and credibility. As it moves into the 1990s, it will continue to grow and to offer opportunity and challenge. Success will depend on how well we respond to three requirements:

1. Because technical communication exists in a social, organizational, and professional context, we must apply systems theory to communication situations. Communication is more than the old sender-message-receiver analogy suggests. It is a system with boundaries, feedback, participants, channels, and interactions.

2. Computerization is taking place at such an accelerated pace that we can no longer stand on the sideline. Word processing is part of writing and editing. Writing is as apt to be online as on paper. Electronic mail and system networking are technical communication delivery systems. Using computers to communicate more quickly and more effectively is a human goal. The recent research of Jeanne Halpern and Sarah Liggett presented in *Computers and Composing*, published by CCCC in 1984, Judy Ramey's award-winning two-part article "Developing a Theoretical Base for On-line Documentation," and the online editing programs being developed by David Farkas are significant contributions and evidence of the growing importance of the computer to technical communication.

3. We must, above all else, avoid becoming isolated from the professional world. Teaching of technical communication must, of course, stand solidly on traditional academic principles and research, but to be valid it must also have

a foot firmly fixed in the professional arena, lest it become unrelated, unrealistic, and archaic.

Whatever its future, technical communication is today a mature and expanding discipline. Our teaching, especially in the degree programs, must have a continuing professional orientation, for the students graduating from these programs will become the technical-publications staff in industry, education, and government. Although writing is their central skill, they must be editors as well. They must understand graphics, illustrations, production, and document design. They must be as prepared for online-documentation and desktop-publishing tasks as for traditional publication tasks. If such is the challenge to tomorrow's students, can the challenge be less for their teachers?

What's Practical about Technical Writing?

CAROLYN R. MILLER

COURSES AND PROGRAMS IN technical writing are both praised and damned for being "practical." Other writing courses are practical, to be sure: in general, practical rhetoric emphasizes that discourse is a means for pursuing a goal. Thus, freshman composition aims to help students be more effective as students, technical writing aims to help them be more effective as engineers or accountants or systems analysts, and the writing instruction that accompanies many literature courses aims to help them to be more effective as reader-critics. But since technical writing is singled out for being practical, it is worth considering what makes it so.

The Meaning of "Practical"

Most immediately, the practical seems to be concerned with getting things done, with efficient and effective action. Furthermore, efficiency and effectiveness seem more important for some types of action than for others; that is, some actions themselves have practical aims (rather than aesthetic or ritual ones), actions concerned with the material necessities of making a living or managing a household. One can thus *be* practical (or impractical) *about* practical action. *Being* practical suggests a certain attitude or mode of learning, an efficiency (or goal-directedness) that relies on rules proved through use rather than on theory, history, experience, or general appreciation. Practical rhetoric therefore seems to concern the instrumental aspect of discourse—its potential for getting things done—and at the same time to invite a how-to, or handbook, method of instruction. Technical writing partakes of both these dimensions of practical rhetoric.

The rhetoric of the early Greeks also involved both dimensions. They emphasized that rhetoric was an art (or techne). This meant (to Aristotle, at least) that rhetoric was conceptualized and teachable (not a knack, as Plato had feared) but neither certain nor absolute (not a science, as Plato had hoped). Greek rhetoric thus initiated both a handbook tradition of instruction and a counterposed theoretical appreciation for the multiplicity of relations between means and ends.

Richard Bernstein has suggested that there are both "low" and "high" senses of "practical," two senses that parallel the handbook and theoretical traditions of rhetoric. It is the low sense, Bernstein says, that calls to mind "some mundane and bread-and-butter activity or character. The practical man is one who is not concerned with theory (even anti-theoretical or anti-intellectual), who knows how to get along in the rough and tumble of the world" (x). The high sense, which derives from the Aristotelian concept of praxis and underlies modern philosophical pragmatism, concerns human conduct in those activities that maintain the life of the community. One of the many reasons for the discrepancy between these two senses of the practical highlights the dilemma of technical writing, which is usually called practical in the low sense (by both its friends and its enemies, incidentally). This reason has to do with the social structure of the Greek city-state, which permitted the free citizen to be concerned with the good of the polis without being much concerned with bread-and-butter activities. The reason, of course, is the institution of slavery. Manual labor and most commercial activity were performed by noncitizens— slaves, foreigners, women. These activities were "preconditions" to the fulfillment of human potential in self-government, according to Nicholas Lobkowicz: "One would almost be tempted to say that the Greeks considered all 'prepolitical' activities prehuman and that only in the political life were they able to see a way of life which transcended the animal realm" (22). Technical writing, the rhetoric of "the world of work," of commerce and production, is thus associated with what were low forms of practice from the beginning. In a world in which it is more dishonorable to own slaves than it is to work for a living, we might question whether this association should prevail.

A Conceptual Contradiction

Before trying to suggest what it might mean to apply the higher sense of practical to technical writing, I want to indicate some difficulties in accepting the low sense uncritically, as many technical writing teachers have. These difficulties are revealed by a contradiction within the self-justifying discourse of technical writing pedagogy: the attempt to hold both that nonacademic rhetorical practices are inadequate (and therefore need improvement through instruction) and that they serve as authoritative models (and therefore define goals for instruction). We seem, that is, uncertain about where to locate norms, about whether the definition of "good writing" is to be derived from academic knowledge or from nonacademic practices. Most teachers will recognize the contradiction in the familiar dilemma of having to admit to students the discrepancy between practices that are supposed to be effective and those that are actually preferred and accepted.

The first side of the contradiction is the familiar justification for teaching technical writing. We teach it because when students graduate and begin writing on the job, they do not do very well. In the technical writing textbook

I use, the first chapter, "Why Study Technical Communication?" documents the "inadequate communication skills of many technical professionals" (Olsen and Huckin 7). For example, it quotes a survey about recently graduated civil engineers showing that writing and speaking are the areas of competence most important to civil-engineering practice but that about two-thirds of recent graduates are judged "inferior" in these areas; results for mechanical and electrical engineers are similar. Complaints about technical writing from senior officials in science and industry include "foggy language," failures of emphasis and coherence, illogical reasoning, poor organization—a familiar litany. Most technical writing textbooks begin with the same rationale, that nonacademic rhetorical practices are wanting. The justification for academic instruction is that academics know something that can help improve professional practices.

The second side of the contradiction derives from the research that interested faculty members have begun to do on rhetorical practices in business, industry, and science. This research is justified not only by the academic assumption that knowledge is a good thing but also (and often primarily) by the belief that knowledge of nonacademic practices is necessary to define goals for teaching practical rhetoric. As Paul Anderson puts it, "We [educators] must first understand the profession, then design our curricula accordingly. Only if we understand intimately the job we intend to prepare our students to perform can we create effective professional programs" ("What Technical" 161).

One of the favorite research projects is the survey, which can show what kinds of work-related writing the population surveyed does, how important it seems to be, what its common problems are, and what qualities and features are valued. In reviewing selected surveys, Elizabeth Tebeaux notes discrepancies between instructional assumptions and industrial practices and concludes that "several curricular changes are clearly mandated" in order to "meet the communication needs of writers in industry" (422). Anderson reviewed fifty surveys, because they can provide "teachers with important insights they can use as they design courses in business, technical, and other forms of career-related writing" ("What Survey" 4). Many surveys, such as those by Marcus Green and Timothy Nolan and by Bill Coggin, have been proffered as authoritative sources of information about what a curriculum should accomplish for its graduates. Ethnographic research has also been justified in instructional terms: according to Stephen Doheny-Farina, for example, "By learning more about nonacademic contexts for writing, we are learning more about the kinds of rhetorical demands faced by many of our college graduates," and this knowledge "can inform the teaching of writing" (159).

Major national grants have gone to researchers engaged in work justified in these same ways, a clue to the institutionalization of this line of reasoning, as well as to its extension from technical writing to composition in general. The Fund for Improvement of Post-Secondary Education (FIPSE) sponsored a project on writing-program evaluation at the University of Texas; the project produced a report saying that "before any college writing program can be judged effective or ineffective, we must know first if what it teaches has value to its graduates in later life. Like any educational program, the overall effectiveness of writing programs must be judged according to the needs of the population they serve" (Faigley et al. 1–2). Another FIPSE grant went to Wayne State for a university-industry collaborative effort on research and curriculum development in professional writing. The researchers present cooperation between academics and practitioners as the way to "ensure that students are prepared for the diverse communication tasks outside the university" (Couture et al. 392–93). FIPSE has also sponsored research on collaborative writing in the workplace by Lisa Ede and Andrea Lunsford, who cite as a major problem "the dichotomy between current models and methods of teaching writing . . . and the actual writing situations students will face upon graduation"; this dichotomy results, in part, from "our lack of detailed understanding about on-the-job writing" ("Research" 69). The National Institute of Education earlier sponsored work by Lee Odell and Dixie Goswami on writing in nonacademic settings; their study also suggests that our ability to teach writing will be "enhanced" by more complete understanding of how people come to write successfully on the job ("Writing" 257).

Practice as Descriptive or Prescriptive

In its eagerness to be useful—to students and their future employers—technical writing has sought a basis in practice, a basis that is problematic. I do not mean to suggest that academics should keep themselves ignorant of nonacademic practices; indeed, much of the research I cited above has been extremely illuminating. But technical writing teachers and curriculum planners should take seriously the problem of how to think about practice. The problem leads one to the complex relation between description and prescription. Odell warns against mistaking one for the other: "we must be careful not to confuse *what is* with *what ought to be*. . . . We have scarcely begun to understand how organizational context relates to writing, and we have almost no information about which aspects of that relationship are helpful to writers and which are harmful" (278). Anderson also warns us about this mistake: in presenting a model of the technical writing profession for use in designing curricula, he cautions that the model "represents an ideal. It is built around the *best* practices of the profession, not around *common* practice—or malpractice" ("What Technical" 165). He gives as examples usability testing (not common but good) and readability formulas (common but bad). Neither Odell nor Anderson, however, gives us much help in understanding what is helpful and what is harmful, what is good practice and what is malpractice. Even David Dobrin's discussion of the contradictions involved in teaching to the standards of employers, although it recommends both curricular and corporate reform, relies finally on accepting practices of the workplace on their own terms; teachers should "make people at work better able to deal with others" ("What's the Purpose" 159).

At this point, it is worth recalling an earlier (unfunded) study of writing in nonacademic settings, "Writing, Out in the World," a chapter of Richard Ohmann's *English in America*. Ohmann avoids the contradiction of taking practice as both imperfect and authoritative by positing a wider perspective from which to make such judgments; he requires, as Odell and Anderson and Dobrin do not, a basis for evaluating a practice other than that of the practice itself. The nonacademic writing Ohmann examined is that of futurists and forecasters, of foreign-policy analysts, and of the government officials who wrote the memorandums we call "The Pentagon Papers." Ohmann sought to establish, not that academic writing is different from writing in the workplace, but that they are dangerously similar: he concludes that academic instruction in writing "has helped, willy nilly, to teach the rhetoric of the bureaucrats and technicians" (205). He claims that the

> writing of the powerful and influential shares some characteristics with the required writing of their college-age sons and daughters; that these characteristics are fairly important to the style of thinking and planning that guides the most powerful country in the world; and that this style has some systematically dangerous features when it operates not in the classroom but on the stages of history. (173)

A similar and more direct charge has been made recently by Susan Wells, who claims that "the ideology of technical writing explicitly assents to its instrumental subordination to capital; the aim of the discipline as a whole is to become a more responsive tool" (247). Being useful is not necessarily good, according to these Marxist critics, but little in the discourse of technical writing allows for this conclusion or explores its consequences. Because the Marxist critique features practical activity as a central concept, it raises questions that are particularly germane to technical writing, questions about whose interests a practice serves and how we decide whose interests should be served.

Practice and Higher Education

The uneasy relation between nonacademic practice and academic instruction has been part of academic discussions about technical writing from their beginnings in the late nineteenth century, as Robert Connors's historical work has shown. Connors documents recurrent debates over whether practical or humanistic goals should prevail in technical writing courses (or, as they were commonly called, "engineering English"), whether, that is, such study should prepare technical students for work or for leisure. Moreover, these debates reflect a larger debate in American higher education, about the appropriate relation between vocational preparation and cultural awareness. In mid-nineteenth century, this debate transformed the American college curriculum, according to the educational historian Frederick Rudolph, who points specifically to the

Morrill Act of 1862 and the founding of Cornell in 1866. The first president of Cornell, Andrew White, "confronted all the choices that had been troubling college authorities: practical or classical studies, old professions or new vocations, pure or applied science, training for culture and character or for jobs" (117). White opted for pluralism, for providing many courses of study in preparation for many kinds of lives: "the Cornell curriculum . . . multiplied truth into truths, a limited few professions into an endless number of new self-respecting ways of moving into the middle class" (119). In a similar vein, Laurence Veysey's study of the emergence of the American university in the nineteenth century traces the development of "utility" as a basis for education. During this period, according to Veysey, "America was a scene of vocational ambition," both in terms of individual aspirations and in terms of the desire for public service. At the same time, the notion of public service broadened to include practical and technical occupations, not just the gentlemanly occupations for which earlier education had been preparatory. "Vocational training," says Veysey, "directly affected the undergraduate curriculum of the new university" (66).

Other commentators have emphasized that the relation between instruction and practice is part of a more general condition, the subsistence of higher education in a socioeconomic matrix. Clark Kerr, in *The Uses of the University*, says that "the life of the universities for a thousand years has been tied into the recognized professions in the surrounding society, and the universities will continue to respond as new professions arise" (111). (This view, of course, implies that the classical curriculum served as preparation not for leisure but for the upper-class vocations of law, politics, and the ministry.) John Kenneth Galbraith has noted that "it is the vanity of educators that they shape the educational system to their preferred image. They may not be without influence, but the decisive force is the economic system" (236). More specifically, in his critique of nonacademic writing, Ohmann comments that

> the constraints upon English from the rest of the university and especially from outside it are strong. . . . [T]he writers of the textbooks and the planners of courses . . . can hardly ignore what passes for intellectual currency in that part of the world where vital decisions are made or what kind of composition succeeds in the terms of that part of the world. (206)

Current enthusiasm for "industry-university collaboration" in applied research and development is perhaps the most recent manifestation of this general and necessary relation. But there is also a repertoire of accepted mechanisms for channeling the relation—internships, advisory councils, certification of graduates, and procedures for justifying and accrediting programs. These mechanisms are used in educational programs for the established professions, like law, medicine, engineering, and teaching, as well as in several areas of practical rhetoric with relatively long curricular histories, like journalism and public relations. For the most part, the channels these mechanisms create are one-way: influence flows primarily from nonacademic practices to the academy. The gradient is reflected in the language at the industry-university interface, which includes, on the one hand, "demand," "need," "value" and, on the other, "response," "service," "utility." My own university, a land-grant institution, provides a case in point. Its "Mission Statement" declares that the university "has responsibility for the academic, research, and public service programs in areas of primary importance to the State's economy." University policies concerning proposals for new degree programs require statements concerning the proposed program's relation to the institutional mission, to student demand, and to "manpower" needs in the state.

Teachers of technical writing have advocated applying the mechanisms of nonacademic influence to their new programs, using the same kinds of language. Internship programs should be adopted in technical communication programs, according to a recent review of literature, because they encourage students to relate their study of theory to practice, permit faculty members to "keep in touch with" current practices, and enable employers "to influence college programs" (Gloe 18–19). Advisory councils are advocated because they "integrate the endeavors of the two worlds [academic and business-industrial] directly and in a[n] . . . effective manner" (Brockmann 137). (Certification has been discussed within the Society for Technical Communication, but there is insufficient consensus in the profession to arrive at standards ["Certification" 6]; accreditation is now being investigated by the society [*Strategic Plan*].)

Such language echoes the discourse of other professional programs, programs that have provided precedents for technical communication.

Library science
It is widely believed and reported that a chasm of mutual ignorance and indifference separates librarians and library educators from one another. . . . All sectors of practice regularly and strongly express a desire for more influence over the content and character of professional education. (Clough and Galvin 2)

Public relations
Practitioners and educators must act in concert to guide public relations in the direction of professionalism. (Commission on Graduate Studies in Public Relations 5)

Information science
Lack of communication between the employers of information professionals and the institutions that educate and train them is one reason that educational institutions are not meeting needs and demands of the changing environment and new technologies. (Griffiths, abstract)

Business
MBA curricula must be reevaluated and, perhaps, restructured if they are to meet business expectations, and—from the point of view of business —if they are to better prepare students for the real world in which they will build their careers. (Jenkins and Reizenstein 24)

Journalism

What training and preparation do radio and television journalists consider important for a career in their field? Answers . . . should contain valuable insights for the broadcast journalism educator. (Fisher 140)

Training and development

Training activities involve a wide variety of skills, abilities, knowledge, and information. . . . An interdisciplinary approach to T&D preparation is important, given the range of competencies required. (Reed 11)

This discourse is infected by the assumptions that what is common practice is useful and what is useful is good. The good that is sought is the good of an existing industry or profession, with existing structures and functions. For the most part, these are tied to private interests, and to the extent that educational programs are based on existing nonacademic practices, they perpetuate and strengthen those private interests—they do indeed make their faculties and their students "more responsive tools." As the minutes of one meeting of the advisory council to the School of Engineering at my university indicate, regular contact between the university and industry "makes students more valuable to industry."

Praxis and Techne

My discussion so far has relied on a set of related oppositions that pervade the discourse of higher education:

theory versus practice
academy versus industry
ivory tower versus marketplace
idle speculation versus vocationalism
inquiry versus action
gentleman-scholar versus technician-dupe
contemplation versus application
general versus particular
knowing-that versus knowing-how
science versus knack

In this form the oppositions are probably unresolvable, and the best we can hope for is Anderson's notion that they should form a "creative tension" (Introd. 6).

Another approach is to suspect the worst: that a dichotomy so widespread must be (at least partly) false. And in fact, Aristotle's characterization of rhetoric as an art, rather than a science or a knack, cuts through these oppositions with a middle term—techne. As he defines it in the *Nicomachean Ethics*, "a productive state that is truly reasoned" (VI, iv), techne requires both particular and general knowledge, both knowing-how and knowing-that; techne is both applicable and conceptualized. Donald Schön's recent critique of professional education relies on the same middle term: it is "art," he says, that professionals display in practice, and it is art that unifies theory and application in a process he calls "reflection-in-action." Aristotle's *techne rhetorike*, or treatise on rhetorical art, joins theory and practice by deriving knowing how from knowing that, prescription from description. Although positivist philosophy claims that this derivation is fallacious ("you can't get 'ought' from 'is' "), one of the major insights of Marx, according to Bernstein, is to deny the positivist fallacy. Marx (as well as Aristotle) is able to derive from description of existing social practices the shape of human need and potential—which provide the basis for prescription.

But to understand Aristotle's *Rhetoric* only as a techne is to miss what Aristotle himself has to say about practice. Understood as techne, Aristotle's treatise would fall within the handbook tradition, as a set of instructions that helps one produce texts. Such a treatise would concern productive knowledge, or *episteme poietike*, one of three kinds of knowledge in Aristotle's system: theoretical (concerned with knowing for its own sake), practical (concerned with doing), and productive (concerned with making). According to George Kennedy, Aristotle does not make the connection between rhetoric and productive knowledge (as he does for poetics) but treats rhetoric as theoretical knowledge concerned with "discovering" the available means of persuasion (63).

The remaining alternative—that Aristotelian rhetoric is practical, rather than theoretical or productive—has been argued by Richard McKeon, and its implications have been explored by Eugene Garver. To see rhetoric as practical, in Aristotle's system, is to emphasize action over knowledge or production; rhetoric becomes a form of conduct, like the related practical realms of ethics and politics, which are constant background presences in the *Rhetoric*. Aristotle distinguishes carefully in the *Nicomachean Ethics* between production and practice, poiesis and praxis: as distinct from "science," or theoretical knowledge, both concern the variable, or that which can be other than it is; but they differ in that production "aims at an end other than itself," the product, and practice aims at its own performance, at "doing well." The reasoning appropriate to production takes the form of techne, art or technique, and the reasoning appropriate to performance, or conduct, takes the form of *phronesis*, prudence; for Aristotle there can be no art, or technical knowledge, of conduct. Prudence is the reasoning that makes one "capable of action in the sphere of human goods" (*NE* 6: v). Like techne, prudential reasoning is situated to undermine the oppositions that plague discussions of professional education, for it necessarily concerns both universals and particulars: it applies knowledge of human goods to particular circumstances (*NE* 6: vii; Garver 64). Unlike techne, however, which is concerned with the useful (that is, with the quality of a product given a set of expectations for it), prudence is concerned with the good (that is, with the quality of the expectations themselves).

Aristotle's concept of praxis has also informed some recent thinking about human action. As the central concept in Marx, praxis highlights the way in which the human person "is the result of his [or her] own work" (Bernstein 39; see also Lobkowicz 418–20). Human belief structures and social relations are understood to be based in practical relations between human beings and objects. Schön's account of professional practice emphasizes the "knowing inherent in intelligent action" (50). Moreover, practices, as Alasdair MacIntyre has insisted, create not only knowledge but their own goods, and because practices are necessarily social, these goods require "subordinating ourselves within the practice in our relationship to other practitioners" (191). The insights for the academic are that practice creates both knowledge and value and that the value thus created comprehends the good of the community in which the practice has a history.

Understanding practical rhetoric as a matter of *conduct* rather than of production, as a matter of arguing in a prudent way toward the good of the community rather than of constructing texts, should provide some new perspectives for teachers of technical writing and developers of courses and programs in technical communication. For example, it provides a reasonable basis for the necessary combination of academic and nonacademic contributions to curriculum. If praxis creates knowledge, academics should indeed know about nonacademic practices. But the academy does not have to be just a receptacle for practices and knowledge created elsewhere. The academy itself is also a set of practices, including those of observation, conceptualization, and instruction —practices that create their own kind of knowledge. Such knowledge allows the academy to provide a standpoint for inquiry into and criticism of nonacademic practices. We ought not, in other words, simply design our courses and curricula to replicate existing practices, taking them for granted and seeking to make them more efficient on their own terms, making our students "more valuable to industry"; we ought instead to question those practices and encourage our students to do so too. Wells's "pedagogy for technical writing" suggests that we should aim "to work within the structures of technical discourse so that students can negotiate their demands but also be aware of the limited but real possibility of moving beyond them" (264). My own earlier sketch of a new pedagogy similarly suggested the need to promote both competence and critical awareness of the implications of competence ("Humanistic" 617). I might now supplement critical awareness with prudential judgment, the ability (and willingness) to take socially responsible action, including symbolic action.

An understanding of practical rhetoric as conduct provides what a techne cannot: a locus for questioning, for criticism, for distinguishing good practice from bad. That locus is not the individual or any particular set of private interests but the human community that is created through conduct; this community is the basis for practice in Bernstein's "high" sense. While the good that praxis in this higher sense creates may include the interests of individuals and industry, it is larger and more complex; the relevant community is not the working group or the corporation but the larger community within which the corporation sells its products, pays taxes, hires employees, lobbies, issues stock, files lawsuits, and is itself held accountable to the law.

Through praxis we make ourselves and each other in interaction: Aristotle emphasizes the political dimension of this interaction, Marx the economic. But whether our everyday activities are primarily those of governing a community or those of making a living, they have both political and economic dimensions. If technical writing is the rhetoric of "the world of work," it is the rhetoric of contemporary praxis. In teaching such rhetoric, then, we acquire a measure of responsibility for political and economic conduct.

LIBRARY
OF
MOUNT ST. MARY'S
COLLEGE
EMMITSBURG, MARYLAND

R ESEARCH ON THE composing process has focused on the author as the originator of a work and as one whose choices, from invention through revision, are guided by the rhetorical elements of subject, purpose, audience, and form. Yet the evidence indicates that textbooks and course syllabi are often at variance with practice: in the corporate setting, the author of a complex document, such as a contract proposal or a technical documentation manual, is rarely considered the originator of or the final *authority* for the document. Corporate writing is thus a cooperative, collaborative effort in which members of different units and departments cross boundaries and work together to produce a document. No one person writes the document; rather, the corporation (or body of individuals that make up the organization) is the author, and it is the corporation that is the final authority for the document.

In this section, the essayists move from a focus on the solitary writer to the collaborative writer in a social-corporate setting, and they explore the role the organization plays in shaping the writing process and the production of a complex corporate document. Using ethnographic methods of inquiry, each essayist examines writing in the corporate setting. Roger A. Grice's observations come from his own experience as a documentation writer for twenty-two years. Mary Beth Debs's are drawn from an intensive study of documentation writers employed by a large producer of software and computers. Jack Selzer's are based on a case study of the writing processes of an engineer who works for a leading design and systems firm and who, as manager of its Chicago office, writes numerous corporate documents.

These three studies of writers at work tell us that, with the exception of memorandums, letters, and, perhaps, brief reports, most corporate documents are composed not by one person drafting and revising but by a cadre of persons working together. Further, the corporate structure and the chronology of product-then-information development influence—if not dictate—the process of composing technical documentation.

As Grice describes it, technical writers are at the center of the process, but they must seek help from others: for example, content from product developers, verification from internal reviewers, graphics from illustrators, printing from production specialists. In addition, technical writers—all working as team members—generally set the objectives and parameters for the document, compose it, and manage all actions that go into its final production. According to Debs, the organization's division of labor creates the need and establishes the system for collaborative writing, the process by which the resources of the organization are marshaled to produce the document.

Clearly, the social context offers the guidance that enables document production; but, as Selzer points out, that same context exerts constraints that "disable" production. The process of producing technical documentation is not always smooth: writers must often begin composing the documentation before all the content is available, content may change, deadlines may be moved up, products may even change. To complicate matters further, collaboration within the writing team and across departmental lines is not always as it should be: personality conflicts, territorial rights, unwillingness to compromise, and lack of a shared sense of responsibility for the documentation may impair the collaborative process. In addition, the inability to manage time, allocate resources, negotiate differences, or reconcile multiple voices impedes progress.

The three essays indicate that the realities of workplace composition are far more varied than current classroom instruction leads students to believe. The essayists suggest that, instead of stressing the solitary efforts of the writer to develop finely tuned prose, our textbooks and courses need to teach students about organizational and group processes—to plan and manage complex organizational writing projects, negotiate with team members, resolve group conflict, and expect and deal effectively with the unexpected.

Document Development in Industry

ROGER A. GRICE

LOOKING FOR A WAY to arouse interest in the first technical communication course I taught to engineering and science students, I let them in on a secret: only a small portion of their careers would be spent doing the work for which they were trained; the rest of the time they would spend communicating with others, either telling others of their work or finding out what others were doing. The approach was apparently successful. The students paid careful attention to lectures and did their assignments promptly and diligently. They were aware of the importance of communicating and working with others in starting and furthering their careers.

What then of our technical communication students—those who are training for careers as professional technical communicators? What words of wisdom do we have for them? I believe the message should be the same for them as for the engineering and science students: only a portion of their time—Lawrence McKinley estimates it to be twenty-five percent—will be spent doing what they trained for (writing technical documents), and the remainder will be spent communicating with and working with others. And as with the engineers and scientists, this communication will consist of telling others of their work and finding out what others are doing.

While we are getting better at teaching our technical communication students about how they will spend twenty-five percent of their careers, we are often lax in teaching the other seventy-five percent: collaboration and organizational communication. Yet the realities of the workplace not only account for a large percentage of a technical communicator's time, but they also help shape the strategies that technical communicators use to do their writing.

A New Look at the
Technical Communication Process

As the production of information becomes a major activity of the American work force, the need to produce this information in the most effective and efficient way becomes of prime importance to our economy. Attention must focus on the methods and processes organizations require to produce and use technical information. This aspect of the communication process has often been overlooked in the past. Instead, studies have focused on the purpose a piece of writing is to serve (Kinneavy), the processes that individual writers follow when creating or revising texts (Emig; Flower), or quantitative measurements of textual characteristics (Gunning; Flesch).

Communication theory, the study of how a message is transmitted from a sender to a receiver in an environment, has been examined by many researchers and theoreticians. George A. Barnett and Carol Hughes provide an extensive review of theories and research that apply to technical communication; they explore messages, sources, receivers, channels, and environments. Researchers have examined the writing processes of students (Emig; Flower; Flower and Hayes, "Cognitive"; Flower, Hayes, and Swarts) and adult writers (Berkenkotter; Selzer, "Composing"), the implications of performing the writing in nonacademic settings (Odell, Goswami, and Herrington, "Studying"; Faigley; Odell; Paradis, Dobrin, and Miller), and the effect of industrial environments on writing strategies (Green and Brooks).

But if the technical communication process is to be judged on its own so that it can be developed and improved, we must come to see it not merely as an "industrial branch" of the writing composition process, and certainly not as simple data transfer (in the sense of transferring data over a telephone wire), but as a process with its own clearly defined steps, goals, and constraints.

In this paper, I present an overview of the process and two particular topics of interest: (1) how collaboration works in the process and (2) how the availability of source material affects writing strategies. I also review the implications for teaching technical communication.

Process Overview

In many ways, the process used to develop technical information must mesh with the process used to develop the products that the technical information describes. The parallelism helps ensure that those who develop information are perceived as organizational equals with those who develop hardware (the engineers) and software (the programmers).

The steps of this information-development process are:

1. Product-review phase.
   Information developers learn about the product and understand the schedules under which they must work.
2. Objectives phase.
   Information developers set objectives for the information they will produce.
3. Specification phase.
   Information developers specify the parameters of the information packages they will produce so those information packages will meet the objectives that have been set.

4. Development (writing) phase.

Information developers write drafts of their information units and arrange for inspecting, testing, and editing those information units.

5. Verification phase.

Information developers verify the accuracy, completeness, and suitability of the information they have produced.

6. Production phase.

Information developers put information into its final form and send it to a printer for printing and distribution.

7. Maintenance phase.

Information developers receive feedback on the information they have produced and make changes in response to that feedback.

8. Quality-assessment phase.

Information developers determine how effectively and how efficiently they worked to produce high-quality information.

Because this information-development process is so closely tied to a technical-development process, it is heavily dependent on factors associated with the technical process, factors that are frequently ignored when discussing a writing process. Two of these factors deserve special mention: the importance of collaborative efforts and the effects of source material on writing strategies.

Collaborative efforts

Producing technical information is a collaborative effort (Arms; Debs; Dressel; Goble; Palokoff). Consequently, any writing strategies adopted by information developers must take the collaborative nature of the process into account. While the creation of a draft (apart from such considerations as source material, review, and production) may be done by one person, creation by one person working alone is the exception rather than the rule. Drafts usually involve the work and contribution of a number of people.

For example, a group of writers may work as a team with a lead writer—perhaps a senior member of the group—coordinating their efforts. In such a situation, the strategies of the individual writers must be compatible with the working of the group as a whole. The need for compatibility does not imply that all writers must necessarily follow the same strategy; it does mean that the strategies adopted must mesh with one another to accomplish the overall task assigned to the group. For example, if the lead writer determines that each writer will produce a "quick and dirty" first draft to distribute among the writers in the group so they will all know what the others are doing, it would not be a workable strategy for one writer to decide to work on individual pieces of a draft, polishing and refining each piece before moving on to the next piece.

Similarly, any realistic strategy must take into account the technical reviewers, graphic designers, editors, test personnel, production specialists, and others with whom the technical communicator must work. To ignore any of these groups and the demands they place on technical communicators or the services they provide to technical communicators is to take an unrealistic view of the job to be done.

One important collaboration required to produce technical documentation is the production, distribution, and use of source material—the information produced by product developers to describe the work they have done to develop the product. Technical communicators use this source material as a basis for the information they produce for customers.

Effects of source material on writing strategies

The gathering and understanding of source material are major activities for most technical communicators; not only does the source material determine the content of the finished piece of technical communication, but it also affects the strategies used to produce it. It is usually necessary for technical communicators to start producing drafts before all, or sometimes most, of the required source material is available (Vreeland). As the source material becomes available in "chunks," technical communicators must factor it into the format that has been selected for the piece of information they are developing. If the source material that becomes available is expected, there may well be a place set aside for it in the outline. If, however, the source material describes a new or unexpected topic, the outline will probably require change to accommodate it. If schedules are not tight, there may be time to integrate the new material thoroughly so that the finished product reflects the organization of all the source material. If, however, there is little time for reorganization or if the source material is made available late in the cycle, it may need to be incorporated hurriedly, with the result that the final version appears patched and less cohesive than it should be.

Another factor to consider is the form of the source material. Unlike academic researchers, technical communicators in industry rely more heavily on people than on documents as sources of information (Vreeland; Hollister; Saar). It would seem that source material obtained from people would be incorporated into documents differently from source material obtained from documents. For example, when source material is obtained from printed documents, technical communicators might well use the strategy of "cleaning up the specs" as a way of incorporating the material. When source material is obtained orally from a product engineer, technical communicators would do better to use the strategy of "translating" the words and perspective of the engineer to those of the reader-user. Similarly, information obtained through testing the product (with the technical communicator observing or participating in the test) would be more apt to be incorporated as procedural information than as narrative.

The need for collaboration and the availability of source material are just two of the factors that affect the strategies used by technical communicators working in industrial environments. Other factors to be considered include standard outlines to structure documents, content guidelines to determine what goes into the documents, checklists to assess completeness and effectiveness, boilerplate to develop drafts, and readability formulas to assess documents.

Implications

The process description in this paper, while short, is rich in implications for those who teach technical communication. This process description could serve as subject matter for a course or a portion of a course so that technical communication students can understand the way information is developed in industrial organizations. It could also serve as a model for writing assignments, wherein students start with source material, set objectives, and work through the process to produce a final document that may require modification and maintenance.

The technical communication process is full and rich. The writing phase is just one portion of the process; the prewriting and postwriting phases constitute a large portion of the total information-development effort and do much to shape the form, content, and effectiveness of the final information product. The prewriting activities, such as gathering requirements and source material, setting objectives, and developing specifications, and the postwriting activities, such as rewriting and revision, verification, collection and analysis of feedback, production and publishing, and maintenance, are as important to the technical communication process as the writing phase. When considering or describing the process, we must consider all the process phases, not just the one with which we are most familiar.

Information development is not a one-person job. From the earliest stages of the process to the end, information developers work with others—product developers, marketing representatives, testing personnel, graphic designers, editors, and production specialists. While technical communicators most frequently work in collaboration with others, many, if not most, classroom assignments stress individual efforts to develop clearly written prose. Collaboration must be strongly factored into academic writing assignments—in the way that assignments are structured and in the way that the profession is described and taught.

Information development is an integral part of product development. Technical communicators must see themselves as professional equals with the hardware developers and software developers with whom they work. And the hardware developers and software developers must see the technical communicators as professional equals. Information developers can, and do, make significant contributions to the products on which they work—both to the information and to the design and development. This sense of professionalism and responsibility must also be built into technical communication curricula.

Activities not always associated with technical communication must be made an integral part of our courses; these activities, often viewed as beyond the scope of our courses, are vital to the success of an information-development effort. The ability to set objectives and to know if and when those objectives have been met is key. The ability to analyze audiences, user tasks, and available media must be mastered before an optimum information package can be developed. Verification of information—by whatever means possible—before it is delivered and analysis of feedback—in whatever form—after it is delivered are critical to the successful development of customer information.

The strategies needed and used by professional technical communicators to develop information are often more wide-ranging and complex than those that are taught. These strategies involve more than writing; they include interpersonal communication skills, the ability to change tactics as products and organizations change during product development, the ability to work with others, and a willingness to work in an environment where the givens are far less numerous and far more subject to change than students may be led to expect.

In sum, technical communication students need to understand the entire information-development process, not just the writing portion of it.

Collaborative Writing in Industry

Mary Beth Debs

COLLABORATIVE WRITING AND coauthorship are not particularly new areas of investigation. Much of the early source material comes from sociology and communications, where studies on collaborative activities and networks of influence began as early as 1955; later studies were closely related to Diane Crane's work on "invisible colleges" and Thomas Kuhn's paradigm theories. These earlier empirical studies documented multiple-authorship trends in the sciences, where seventeen authors listed on a single paper is not uncommon (Clarke; Krohn; Price and Beaver; Zuckerman); they documented the importance of communication, particularly oral but also written, in scientific and technological organizations;[1] and they attempted to document the ways in which team research and group authorship were mechanisms for the socialization of new members into an organization or discipline (Blau; Cole; Hagstrom; Lodahl and Gordon; Merton; Weinberg). Given this particularly rich sequence of studies, it is somewhat surprising but certainly welcome to find within the past five years researchers in the fields of composition and technical communication investigating the range of social interactions that make up writing processes within industry and focusing particularly on collaborative activities.

Much of the impetus for this focus comes from a change in research location and a necessarily concomitant shift in theoretical perspective for many researchers in composition and rhetoric. Led to some extent by a recognition in writing-across-the-curriculum programs that little specific information existed about writing outside the classroom, researchers such as Lee Odell, Dixie Goswami, and Anne Herrington initiated studies of writers and writing in nonacademic settings. At the same time, a framework for exploring the cultural or social dimensions of writing was taking shape, informed largely by work in sociolinguistics, cultural anthropology, and social psychology. Because most writers in nonacademic settings are writing as members of a particular social unit (the organization in which they are employed), this social perspective is essential to understanding the relations between organized social activities (activities that make up the context and create the exigence for writing; see Harrison) and the production of written communication. As Lester Faigley notes in his review of the foundations of social theories of discourse, "The social perspective, then, moves beyond the traditional rhetorical concern for audience, forcing researchers to consider issues such as social roles, group purposes, communal organization, ideology, and finally theories of culture" ("Nonacademic" 236). Studying writing as it occurs within the organization brings under investigation the individual's composing process, the set of interactions that take place during the production of a document, and the contributory nature of those interactions as well. Examining the role of collaborative writing in industry, we come to study the processes of writing and of organizing.

This essay, then, has two aims. First, it offers a general review of the concepts that guide research on collaborative writing, the issues that such research raises, and the factors that seem to influence collaboration. The studies were selected in particular because of the potential relevance of their findings to understanding the complexity of social interactions and the role of collaboration in the practice of technical communication. Second, this essay establishes a basis for incorporating collaborative activities in the classroom and suggests ways to shape these activities so that students may develop the skills needed to manage more effectively the document-production process.

The Extent of Collaboration in the Workplace

Several surveys show that collaborative writing is a widespread activity in industry. Lester Faigley and Thomas Miller report that of the two hundred people responding to their survey, 73.5% write collaboratively at work "sometimes." They note that this collaboration occurs in several different ways:

> Sometimes a half dozen or more experts in various fields will contribute a section to a technical report, with the project leader integrating the sections into a coherent whole. In other cases a superior will simply review the work of a subordinate, making small changes if necessary. And on still other occasions people will work closely throughout all phases of a writing project, coming up with ideas and putting them on paper as a team. (567)

In "What Survey Research Tells about Writing at Work," Paul Anderson reviews the results of six surveys that specifically considered some form of collaboration, including writing with other people, critiquing drafts, and delegating written work. While these studies support the conclusion that collaborative writing in its many guises is widely practiced in the workplace, their results demonstrate that the amount of time individuals spend in collaborative writing activities varies considerably. Research has yet to correlate these variations with any specific factors, such as job title, type of writing, nature of the organization, or organizational structure.

To date, the most extensive survey that focuses exclusively on collaboration has been conducted by Andrea Lunsford and Lisa Ede, as part of a multiple-stage project that will also include detailed interviews. The 520 members from six professional organizations (including the Society for Technical Communication) who responded to the survey indicated that they spend only 19% of their writing time working with another person or as part of a group. Ultimately, Lunsford and Ede (who prefer the term *group* or *coauthorship* to *collaborative writing*) could not justify this low percentage without calling into question modern notions of authorship:

> A full 60% of these respondents (who had indicated they spent more than 75% of their time writing alone), however, contradicted themselves by answering a series of later questions in ways which revealed that they often wrote as part of a team or group. . . . Respondents think of writing almost exclusively as writing "alone" when in fact, they are most often collaborating on the mental and procedural activities which precede and co-occur with the act of writing, as well as on the construction of the text. (73)

Nonetheless, the majority of the respondents find writing as part of a team or group to be productive when compared to writing alone.

Defining Collaborative Writing

These surveys uncover three of the difficulties of conducting research on writing in industry. When respondents say they write alone, they are often demonstrating a generally held view of the writing process, albeit an unfortunate one, that narrows it to the physical activity of composing (with pen in hand or word processor at desk). In addition, their perspective may reflect an appreciation more traditionally given to the solitary nature of authorship. Examining personal influences in the literary world, Harold Bloom labels one aspect of this concern "the anxiety of influence." Moreover, and this is a major difficulty with the study of the type of collaborative writing that occurs in industry, there is no consensus (certainly no operational definition) as to what collaborative writing consists of, or what separates a collaborative activity from a merely cooperative one, or under what conditions a series of interactions makes up a collaborative writing process.

Although surveys, interviews, and field studies indicate that industry allows—perhaps, more accurately, encourages—all the variations of collaborative writing that we find in the classroom and adds several more types, researchers have only begun to catalog the various activities that may define the category of collaborative writing. Applied to activities in the classroom, collaborative writing traditionally meant group invention or revision (Bruffee; Clifford). Thus it was (and is) most often used as a way to organize the classroom activity to provide a kind of rhetorical feedback to students (Gebhardt). In current practice, especially in technical communication courses and in writing-intensive courses in engineering, students often write as members of teams, each member equally sharing responsibility for the final product. This group participation is also a pattern relatively common to industry, but one that can be played out in various ways, depending on the division and arrangement of tasks in the writing process. Assume, for example, that the writing process can be conveniently divided into three stages: (1) invention or planning, (2) writing, (3) revising or editing. Recognizing that groups may write a draft together or may separate at different stages and assign individuals to different tasks, we can anticipate at least ten patterns for organizing this type of collaboration: for example, a group may plan, write, and revise together; one person may write, another edit; the group plans together, but parts are written by individuals; and so on. In fact, Lunsford and Ede asked respondents in their survey to indicate how frequently they used seven of these patterns. No single pattern stood out as a first choice, but the results suggest that procedures in which a draft is revised by a person who does not consult with the writer or team are rarely used.

The terms *coauthorship* and *group authorship* seem no less problematic. Immediately, we run into practices in the sciences where three, six, or seventeen authors are listed but where, in some cases, only one person drafted the article. One may argue, however, that here collaboration is part of rhetorical invention: even when only one researcher from a team assumes responsibility for producing the final document, the text will reflect a number of rhetorical choices and considerations already anticipated and negotiated with other members. In addition, participation of both writer and audience within one field will render some choices conventional. This type of authorship can be contrasted to a practice in industry in which writing is delegated (and the writer may draft the document as if he or she were collaborating with the "author") but the writer's rather major contribution is not acknowledged in any formal way. The increasing use of "boilerplate" is another irregular case, since more than one author contributes to a text, although no personal interaction may occur.

Consistently we find in industry that the production of any text is a social process. The activities of this social process are based on the day-to-day ongoing talk embedded in the context; thus the writing process is often "submerged," to use a term suggested by James Paradis, David Dobrin, and Richard Miller, who also describe another type of collaboration which they label "draft or document cycling."

As one form of communication, a form that leaves a permanent record, writing is used by the organization as a means to several ends. First, writing is a product—in the sense that it is something going out ("leading forth") from the organization. Writing also facilitates the organization's operation inasmuch as memorandums, letters, instructions, or reports are used to maintain the context in which they occur. (Lunsford and Ede's results suggest that shorter internal documents, such as memorandums and letters, are likely to be written by an individual, while longer documents such as proposals and reports, which we can speculate are in many cases addressed to external audiences, are likely to be written with at least one other person.) The organization, not the writer, is the final authority for any text it sponsors—particularly those texts addressed to audiences outside the organization. As if to illustrate this authority, there is an increasing trend in industry to acknowledge only internally writers and other contributors of certain documents, such as manuals and company reports.

The division of labor within organizations sets up the potential for collaboration. Members of the groups most closely associated with the design of a product, for example, provide information to the writer. Then, as some members of the organization are designated editors or reviewers, the division of labor extends into the "typical" writing process. The writer has room to negotiate agreements with members authorized by the organization to intervene in the writing—but negotiate he or she must, because the writer's role as spokesperson for the organization can be realized only through collaboration (an interaction leading to a consensus) with other members. The existing context of a company or other organization (created because members have agreed to work together) and the existing division of labor (specialized roles) are two primary features that distinguish collaborative writing in the organizational context from the prototype of two or more authors agreeing to write together.

My purpose is not to make the concept of collaborative writing needlessly complicated but to point out complications that already exist and that affect the direction of research and our ability to interpret reliably and use the results. Surveys, though useful, are a limited tool in exploring the complexity of social systems within organizations. Qualitative methods—particularly variations of discourse-based interviews, account analyses, and observations conducted over substantial time periods—are more likely to yield the detail needed to delineate and describe the function and success of different types of interactions in the writing process. As this data base develops, the term collaborative writing may continue to be applied to any point on a continuum that has at one pole a fully realized version of group authorship and at the other pole a single interaction in the writing process (thus grading a paper or even changing a semicolon could be seen as a form of collaboration).

Small-Group Theory and Collaborative Writing

Several recent studies, using primarily qualitative methods, have examined the group processes involved in collaborative writing. They have focused on well-defined groups and have isolated interactions that are part of the writing process from those that may occur in a broader context. Although we need to be cautious in generalizing from their conclusions since some may be applicable only to the groups studied, these studies are particularly promising. Informed by theories of small-group behavior and interpersonal communication, the research begins to describe in some detail features that may characterize collaborative interactions.

In their interview-based study of twenty experienced collaborators (the group included teachers as well as other professionals), Nancy Allen et al. have determined three features that distinguish "shared-document collaboration" from other types of collaboration: (1) production of a shared document, (2) substantive interaction among members, and (3) shared decision-making power and responsibility for the document.

Writers of shared documents negotiate with other members of the group choices a writer would typically make alone; thus, much of the interaction is devoted to this joint decision making. Allen et al. also contend that divergence and, in some cases, specialization bring about collaboration: in other words, groups are formed to take advantage of different members' expertise or to divide a project in order to save time.

Stephen Doheny-Farina's ethnographic study of a group of computer-software executives focuses on their collaborative activities in producing a business plan for their company. Although detailing group behavior, the study illustrates the reciprocity of influence that exists between rhetorical activities and the organizational context: the collaborative process highlights the conflicting interests of the group members, and it contributes to changes in the company structure and lines of authority. Interestingly, members resolve these conflicts, in part, by debating rhetorical choices and ultimately agreeing on an interpretation of the rhetorical situation.

Janis Foreman and Patricia Katsky's investigation of the problems students encounter during an academic group-writing project in the MBA program at UCLA should be included here because it summarizes characteristics of successful collaboration suggested by Allen, Lunsford and Ede, and other researchers. The authors argue that the well-written report "represents the team's successful working through of *both* small group and writing problems" (23). Three problems in small-group processes that may occur throughout the writing process are poor conflict management, personality differences or unresolved emotional issues, and poor client management.

Foreman and Katsky suggest that, early in the project, students should identify each other's capabilities, define the task, and structure the group as needed. Other problems, which may show up later in the collaborative process, include poor division of labor and procedures for assessing progress and difficulties in reaching effective compromises. The authors identify as problems in the writing process failure to determine the rhetorical situation, to allow time for revision, or to resolve multiple voices in the report.

Collaborative Writing in Technical Communication

None of the research cited above deals with technical communicators exclusively, but it is likely they share with other professionals many of the same assumptions concerning authorship and collaboration. The need for "good" collaboration, for example, underlies most articles that describe or prescribe the document-design process, though it is rarely dealt with explicitly: "The writer and editor must work together closely. . . ." "When the editor and writers are satisfied with the results. . . ." "The project coordinator establishes schedules for completing each phase of the project, including all documentation."

As part of a field study to investigate the role of collaboration in technical communication, I interviewed twenty-four members from a group of technical writers in a software-development and computer-manufacturing company.[2] The majority of these writers at first denied—sometimes vehemently—that they engage in collaborative writing; later, after they had described the discussions and negotiations that go into the choices they make while writing different manuals, they concluded, often with a statement of surprise or reconsideration, that they do collaborate. In one case, of the eighty text features discussed during the interview (including questions of audience, style, and layout), sixty had been determined by the writer during discussions with eighteen other members of the organization, yet initially this writer too had responded that she rarely collaborates.

As we have seen, two or more people writing together is clearly the general conception of collaborative writing, but the activities that accompany the process of writing within an organization sponsoring the writing are also a form of collaborative writing—or, at the least, these social interactions have the potential of being collaborative. That both writers and observers have difficulty in recognizing the potential seems to be a consequence of the organizational context. For the technical communicator, collaboration and the procedures for producing a document are "structured into" the existing organization; that is, they are part of the social structure, the system for getting things done internally.

Technical communicators may work within relatively limited systems in which they interact primarily with one other person or one group: as a writer for an engineering consulting firm, for example. More typically, they work in complex organizations made up of groups, units, divisions. Outlining the process used to produce a technical document in a complex organization, Roger Grice (in this volume) shows that the documentation writer interacts with members from at least four groups: other writers, print-production specialists, technical personnel, and managers. In other types of organizations, the list may include scientists, programmers, engineers, users or audience, and members of other departments, such as public relations, legal, marketing, accounting, and manufacturing. Each division or group represents a community that has been established through the function of what Kenneth Burke calls "identification" (a person may also identify with more than one group within the organization).

Clayton Aldefer and Ken Smith argue that intergroup boundaries (physical and psychological), differences in the types of resources available, and affective patterns (the association of positive or negative feelings with group membership) shape the relations between groups within an organization. In addition, they note that "groups tend to develop their own language (or elements of language, including social categories)" (40). By approaching writing as a collaborative process, technical communicators cross these boundaries while taking advantage of the diversity of viewpoints, but their interactions with members of other groups will be influenced by a variety of factors. Not all interactions are collaborative. We can anticipate that the stronger a negative factor or the greater the number of negative factors, the more difficult it will be to develop a collaborative relationship. In a complex organizational system where the writer must obtain technical information from more than fifty sources, this part of the writing process may bear little resemblance to small-group processes.

Based on responses in the interviews I conducted, the most intensive and problematic interactions occur between writers and members of the technical group—the engineers or programmers responsible for the product. In some cases, difficulties emerged because the writer or members of the technical group did not have a sense of shared responsibility for products or were not encouraged to share responsibility. The problem may be a consequence of different group memberships, different perceptions of the status of the documentation and the product, or different specializations, but the technical writer is in a unique and often difficult position. The technical writer is often obligated to a preexisting social reality—essentially, the knowledge base invented along with the product. The ways of speaking identified with the product—and possibly some of the language used to describe the product—originate with the designers of the product. To collaborate with members of any group, a writer must be conversant with that group—this is a condition of participation. Paradoxically, it is also a threat to the writer's strongest position—that of advocate for the audience external to the group. We can see this situation most frequently in computer documentation, in which the aim often is not to converse with the audience (an aim reserved instead for marketing publications) but to make the audience conversant with the product's developers. The technical writer's role as translator from engineers and programmers to user and, in some cases, from user to programmers, engineers, and managers is not always understood by the groups involved. Collaboration, then, can be used internally as a way of increasing the exchange of information and improving channels of innovation. In addition, collaboration provides opportunities for the writer to contribute to the complete design and development process.

The division of labor—assigning roles according to specializations and orienting groups to particular products—characteristic particularly of large organizations creates the potential (and need) for collaboration. And again, the writers interviewed indicated that the frequency of interactions increases when projects are new, ambiguously defined, or in any way outside the usual practices of the organization. A new product requires the writer to deal with a larger proportion of changing information about the product. With new manuals, much of the interaction seems to cluster near the beginning of the project during the planning stages and, again, near the end.

Finally, in technical communication, collaboration (particularly shared-document collaboration) seems to play a specific role in introducing new writers to the organization's procedures and standards. Jean Lutz points out that the ten interns in her study were assigned to a mentor, "and all of their documents, particularly during the early stages of their tenure with the organization, were reviewed by these supervisors." Apprenticeships for interns and new employees may also take the form of coauthorship.

Collaborative Writing in the
Technical Communications Classroom

Experience suggests that students benefit only a little, if at all, when collaborative writing activities are introduced simply as simulations of "real world" processes. This review of research on collaborative writing demonstrates that, as part of these activities, we need to teach technical communication students to anticipate organizational and group processes, to identify and control variables that affect interactions in the workplace, and to develop strategies for accommodating the intervention of other people within their own writing processes. Each in-class activity should be designed to develop particular skills in organizing.

Initial assignments should be presented in some detail and should focus on exploring why people write in groups: combining resources, saving time, incorporating different perspectives, having an audience at hand. For an early short assignment, groups may be asked to try out different ways of organizing: one group writes together; another group writes parts separately but revises together; another plans together and then separates the writing into parts. There are other combinations; the trade-offs between the different methods can then be compared. In other assignments, students may be asked to represent different interests of a sponsoring organization. Discussions may also consider the collaborative possibilities of editing tasks.

Students should also gain experience managing and analyzing projects that require collaboration over an extended time period. They should begin longer group projects by defining the tasks at hand, evaluating their own abilities to contribute, and reaching a consensus about the rhetorical situation of the project. Students should also recognize that some tasks (such as planning) are often best done by the group, while others may be accomplished better by an individual (e.g., the penultimate revision). Groups may then develop and submit a formal management plan that includes division of the tasks, deadlines, time line, meeting schedule, organizational chart, résumés, and other materials that focus on the process of group organization and process management. For long-term projects, frequent meetings to exchange information and renegotiate decisions must be scheduled. Requiring regular progress reports will help identify and solve problems before they can seriously delay the projects.

Recently, I have begun asking students in my advanced technical writing courses to keep journals or logs recording their group interactions. At the end of the project, I ask all the members of a group to submit their reports to me, summarizing their journal entries and responding to a series of questions that evaluate the performance of the group. These reports make evaluating individual contributions much easier. More important, they again focus the student's attention on the process of organizing and the demands of collaborative writing. To complete the report, the student must consider specific factors that have come up in the abstract during class discussions and now have contributed to the group's success or difficulties.

Recent research indicates that a person's writing process within an organization is often collaborative, though the collaboration may take many different forms. We also are fairly certain that the technical communicator is regularly involved in collaborative activities. In industry, the choices a writer makes are often drawn from a rhetorical situation that is determined and mediated by the organizational context. Other members of the organization may be authorized to intervene in the writing process and negotiate specific text features. Certainly they provide needed information to the writer. The writer, as a member of the organization, must also be aware of his or her role as spokesperson for the organization and can best discover the dimensions of this role through successful collaboration with others.

This review offers a tentative guide for analyzing the types of collaboration that occur during the production of a document and for identifying the strategies that enhance collaboration. It also suggests a number of questions the answers to which could contribute in significant ways to our understanding and our control of certain practices in technical communication. For example, is there evidence of a "corporate hand" in documentation as well as in other written products of an organization? How do writers learn this ethos, and what ways are available to them to change it? Do those who intervene in the writing of a document have the same responsibilities as a coauthor would? Or the same as an audience? And if collaborative writing is a process of organizing, as this essay suggests, then the question we have yet to ask is, "Are there ways that this process ought to be organized?" We are nowhere near answering that question, but it is imperative to note that current research on collaboration and other types of social interactions has made not only the question but also the answer possible.

NOTES

[1] For a comprehensive review of these articles, see Bazerman, "Scientific"; Miller reviews a number of these works in her 1985 bibliographic essay, "Invention in Technical and Scientific Discourse."

[2] The field study used a combination of methods: interviews of 15 writers and 3 supervisors, discourse-based account interviews of 7 writers, and feature analysis of 25 manuals; see Debs, "Collaborative," for a complete report of the research and discussion of methods used.

Composing Processes for Technical Discourse

JACK SELZER

IS IT AN EXAGGERATION to claim that technical writing courses usually approach the writing process in less substantial and less sophisticated ways than typical composition courses? Research on composing has only superficially influenced the conduct of courses in technical writing, in part because until recently nearly all studies of composing considered not the work of practicing engineers or scientists or managers or professional technical writers but the poems, novels, and essays of professional writers or students (Selzer, "Exploring"; Warnock). To put the matter another way, nearly every study of the composing process has regarded writing as a private or personal act of making meaning, as a self-contained process that evolves out of the relation between writers and their emerging texts—and not as a public and social act of exploring ideas and transacting business. Researchers have typically, though not exclusively, observed professional and student writers working alone on essays, fiction, and other projects closely associated with belles lettres; rarely have they investigated how people working cooperatively on technical and scientific enterprises produce reports, manuals, correspondence, and other documents within complex organizations.

I do not mean to say that research on composing has been totally useless for teachers of technical writing. No one who has read the work of Donald Murray, Janet Emig, or Nancy Sommers can ignore the implications even for documents as highly transactional as résumés and reports, although all three tend to see writing as a process beginning with the impulse to put words on the page (instead of as a process compelled by a writer's relation to a community). Nor can anyone ignore the contributions of Linda Flower and John Hayes, even though they confine their studies to the laboratory and seem most interested in cognitive processes that arise prior to any social and rhetorical influences, because Flower and Hayes have obtained highly detailed accounts of composing activities that are instructive for every writing teacher. Such research, however, has only lately begun to consider how meaning emerges not only out of an individual consciousness engaged in private exploration but also out of social interchange—how (in other words) meaning emerges out of many messages strung together into communication chains that are created or encouraged by complex technical and business organizations.

The purpose of this essay is to capitalize on the recent research on composing in the workplace. In the past five years several researchers, especially by using techniques derived from ethnography, have uncovered strategies and tactics employed by scientists and engineers in constructing documents at work. In particular, I mean to survey two related conclusions that have grown out of that work: that composing is a rhetorical act shaped by rhetorical influences and that it is a broadly social act influenced by the culture in which it occurs. I will also sketch out some pedagogical implications of those conclusions.

The Technical Writing Process Is a Rhetorical Act

During the past few years I have observed fairly carefully the writing of a transportation engineer who manages the Chicago office of a large engineering firm (Selzer, "Composing," "Exploring;" Miller and Selzer). I have not been alone in such studies, of course; in fact, I have learned most of what I know about the content and conduct of writing at work by following the efforts of C. H. Knoblauch; John Gould; Lester Faigley and Thomas Miller; James Paradis, David Dobrin, and Richard Miller; Glenn Broadhead and Richard Freed; Janice Redish and her coworkers at the Document Design Center (Goswami et al.); and Lee Odell and his colleagues. In any case, whenever I look back at my experiences with the engineer, I am amazed by the range of writing activities that engage him and at his resourcefulness as a writer. In any given month he works on proposals, various types of correspondence to individuals and groups both within and without the organization, several kinds of reports (both internal and external), professional articles, oral presentations reinforced by supporting documents, and more.

More to the point, he employs a great variety of composing strategies and tactics. He works alone and with others (bosses, colleagues both inside and outside his company, secretaries, professional technical writers, and various combinations thereof). He composes many documents himself, sometimes at one sitting and sometimes (even when he writes documents that are relatively brief) over a period of months. He contributes segments to documents primarily written by others and assembles reports whose parts might be written by several coworkers. He dictates a few things; he composes others in a "first-time-final" manner and submits them to a typist (he rarely types himself, nor does he use word-processing equipment, though some of his coworkers do); he reuses boilerplate from one document to fashion others; and he composes some things painstakingly and laboriously, with multiple drafts and elaborate review procedures. Unlike some of the engineers who work with and for him, who struggle to meet deadlines, seem uncomfortable writing with others, or try to avoid certain writing tasks, my friend produces an astounding amount of capable (if seldom eloquent) prose each month. And while it would be inaccurate and patronizing to say that he does it all effortlessly, that is the impression he sometimes leaves.

What guides his choices of composing activities? A great many are shaped by rhetorical considerations—by the subject he is concerned with, by the audience he is addressing, by the genre involved, and by his own aims and personal quirks. For example, if he is writing a report on a subject well known to him, he might invent by brainstorming some ideas or borrowing from previously written reports; if the issue is new to him, though, he is more likely to consult someone for advice—someone from the network of coworkers and colleagues that he has assembled over the years—or to do some reading on the subject (sometimes at Northwestern University's library). If he is unsure of his subject, he will even on occasion ask a professional technical writer or colleague to draft a statement or construct a figure for him, and then crib from that. Audience also affects his invention tactics: if he knows his readers well, he tends to recall and reuse successful approaches from the past (again with the help of company files), or he modifies consciously those maneuvers that did not work so well. If his readers are less familiar to him, he tries to learn about them by speaking to them directly, either in person or on the phone, and by asking colleagues for information on them. The genres that my friend produces also shape invention: he knows quite well the conventions of the types of documents he produces; he seeks information for professional papers (for example) in ways very different from the ways he seeks information for reports; and if he finds himself writing something for the first time, he schools himself by observing examples of that form written by others. Finally, personal considerations help determine his approach to invention. Is the task important? Is he confident about it? Does he have enough time to work on it? These and many other rhetorical considerations help determine which invention tactics will be employed—and for how long—by one engineer on any given working day.

And it is not just a matter of invention. Arrangement also depends on the subject, on the genre (many documents generally follow one conventional arrangement or another), on the audience (e.g., if a Request for Proposal asks for certain things, the engineer I know will not only include them but will include them in the same order they are listed in the RFP), and on personal habits (e.g., he might put some things before others because that will make the writing easier; unlike many people he outlines in some way nearly everything he writes). Revision is marked by the same considerations. Like the writer known as "Franklin" studied by Broadhead and Freed (74), my friend typically revises superficially and at a single, final "stage" of the writing process. Yet an especially important document might occasionally call forth an elaborate review (involving several colleagues and/or professional technical writer-editors) and substantial revisions, while an unimportant or routine document might never be revised at all. As my friend says, "It all depends on the situation."

The Technical Writing Process Is a Social Act

It also depends on the environment.

In the past half dozen years, building on the work of Bakhtin, Foucault, Vygotsky, and other theorists, many rhetoricians have begun to study the social setting out of which writing—even at its most personal and least transactional—necessarily occurs. Patricia Bizzell, Charles Bazerman, Lester Faigley, Greg Myers, and several others have directed attention beyond narrowly rhetorical concerns (audience, aim, genre) to the ways particular communities influence discourse. Their interest has been in teasing out through careful analysis the intricate and subtle and fascinating ways that settings shape language and in directing teachers and students to note how both the interpretation and the production of discourse require an understanding of the social contexts in which the discourse is created and used.

Whether based on the techniques of literary criticism or ethnography, research into the dynamics of discourse communities naturally interests those who study scientific and technical discourse, for documents written by individuals and groups at work obviously reflect in powerful ways the conventions and expectations not just of their readers but of the particular communities that generate those documents. Thus Bazerman and Myers have explored the social aspects of scientific discourse. Carolyn Miller ("Technology") has compared and contrasted the rhetoric characteristic of scientific prose and technical prose and has also (in her dissertation and in "Genre") shown how the conventions of environmental-impact statements owe as much to the legal and political climate that created them as to the rhetorical situations they are supposed to address. Odell and his colleagues ("Studying Writing") studied how a particular government office served as a discourse community. Paradis, Dobrin, and Miller have discussed the functions of writing within a research-and-development operation. Carolyn Miller and I have cataloged a number of the disciplinary and institutional forces present in reports and proposals by transportation engineers. Louise Dunlap has drawn attention to the ways that an institution's management structure influences communication. And J. C. Mathes has devised a model of some of the social factors typically at work in technical organizations. In short, we have a fair start at understanding how scientific and technical enterprises shape features of the written products created there.

It should not be forgotten, however, that discourse communities also shape the composing process. Compare, for example, the writing processes that an engineer might follow if he or she works for a university with the processes the same engineer might follow in a consulting firm. Even though the two environments might encourage the production of some of the same documents on the same subjects written for the same readers, the writer goes about that production in different ways. An engineering firm might supply the help of a professional technical writer but not a library, while the university might supply just the opposite resources. The engineer in a firm is likely to depend on advisers relatively close at hand; the engineer in the university might well have a more extended network of colleagues to rely on. The engineer in a firm must adapt to one set of external constraints (project deadlines, page limits, budget limitations), while one in a university must adapt to a different set. The rhythms of work in an engineering firm offer certain occasions and opportunities to write, while the rhythms of the school year present other occasions and opportunities. The engineer in a firm depends on one kind of internal review cycle for documents, usually confines revision to stylistic matters (Broadhead and Freed), often works as part of a team, and might subordinate personal efforts to the goals of the group. The same engineer on campus depends on a different (and usually less formal) review cycle, revises more substantially and throughout the writing process, works alone more often, and may well conform less to the group enterprise.

Beyond this, the composing processes of the engineer in either circumstance will be affected by the nature of the enterprise (scientists and engineers differ in the ways they pursue knowledge, and engineers in different fields, firms, and jobs invent in different ways); by one's position in a hierarchy (Paradis, Dobrin, and Miller have shown how writing activities vary markedly as one moves from a staff position to supervisory and management levels); by the leadership of the organization (who assigns work? who sets standards? who enforces work habits?); by the availability of technology (typewriters, word processors, dictation machinery, teleconferencing and electronic-mail equipment, printing and production facilities, etc.: see Halpern and Liggett); and by a host of other local matters, both large and small, that influence writing. The work environment of every technical writer—whether he or she is a technical professional who writes or a professional technical writer—obviously socializes writing activities. Every social context obviously offers freedoms and constraints that enable and disable the production of discourse.

But there is also another, and more subtle, sense in which composing is social. After all, the writing process is not a "procedure" or set of broad activities that can easily be divorced from products. If the writing process may be defined as the way a writer interacts with a developing text in the course of its construction and if the text is social in nature, then composing is also narrowly social in determining how writers conceive of their growing texts. In other words, if social context determines the broad procedures writers follow as they write, it also shapes the kinds of arguments that are offered, the rules for presenting evidence, particular arrangement patterns, even sentence formulas and stylistic conventions. From this perspective, invention grows out of the interchange among writers, texts, and communities; and invention in technical writing means adapting arguments not simply according to the needs of a rhetorical situation but also according to the institution that sponsors the document. Arrangement, in part, means learning and recapitulating the organizational patterns that are current in a discipline or institution. Revision, in part, means choosing a style that matches an organization's ethos (and that may even be codified in a company style manual). As Carolyn Miller has argued

("Humanistic"), technical writing in general might be seen as "a kind of enculturation": a process of learning how to belong to and participate in a community.

Pedagogical Implications

What does all this research imply for the classroom? Most generally it implies that technical writing instructors should pattern their advice about composing not after the practices of professional writers working alone to produce belletristic stories, essays, and poems but after the activities of writers who produce documents in technical, scientific, and corporate settings. In those environments successful writers follow no single, routinized composing process. Rather, they adapt their writing activities to fit specific rhetorical circumstances and to satisfy the social contingencies that emerge from specific organizations and disciplines.

More specifically, then, recent research on composing implies that one job of the technical writing instructor is to produce flexible writers capable of adapting their composing activities to a variety of situations. On the one hand, instructors can elaborate for their students an idealized composing process (complete with ample time for planning, invention, arrangement, drafts, and revisions) that might be employed under ideal circumstances: after all, technical writers need to know how to perfect their prose when the situation calls for excellence and refinement. But on the other hand, since opportunities are rare for writers at work to exercise such elegant and ideal processes, instructors must also equip students with quicker and "dirtier" tactics necessary for writers to thrive in a corporate culture.

In sum, instructors of technical writing should be careful whenever they intervene in their students' writing processes. Unless they wish to arm their students to handle only the narrow range of composing assignments that occur in schoolrooms, instructors will have to resist imposing particular invention and revision tactics on every classroom assignment. For while the idealized composing model that textbooks recommend might suit term papers, masters' theses, magazine articles, or poems, the model is hardly appropriate for most technical prose. Instructors have to prepare students to change tactics when audiences change, when deadlines are short, when the importance of a document waxes or wanes, when the students change jobs, and when other people cooperate in the writing, and so forth. Instructors must expand students' repertoire of composing options so that students can invent, arrange, and revise in many ways, depending on circumstances. Flexibility and choice must be the keys to instruction: just as a teacher of style seeks to stretch the stylistic repertoire of students so they can adapt to a number of rhetorical situations, so too a teacher of composing should strive to expand the range of composing activities available to writers whether they are writing a proposal or report, a routine memo or an important one, with others or alone.

All these things can indeed be taught. The range and richness of a technical writer's composing activities can be expanded through a combination of direct instruction, practice, and evaluation. The direct instruction might come from lectures or from reading. Flower's *Problem-Solving Strategies for Writing* is particularly useful in suggesting new composing strategies, for instance, and there is a growing body of testimony about composing by technicians and scientists that both instructors and students may draw from. Or students can study directly the composing strategies that work in particular situations. Such studies can be relatively informal: Dixie Goswami and her colleagues (17–27) suggest that students keep logs or journals of their writing activities during a semester; when students compare one another's logs, they consciously note the rhetorical dimension that governs composing activities. Or those studies can be more formal: I begin my own technical writing course for scientists and engineers, for example, by assigning students to report on composing activities that they observe (by using ethnographic methods) in particular organizations; the ostensible purpose is to get students to teach me about the dynamics of composing at work, but of course the assignment enriches students' understanding as well.

Other assignments can also teach composing, especially by forcing students to practice new tactics. One assignment might incidentally force students to learn how to use library resources, for example; another might force students to learn how to interview, or to explore the conventions of an unfamiliar genre (e.g., the résumé or proposal), or to study the kinds of arguments and arrangements likely to be productive in technical reports in each student's field. Still others might require students to adapt material used in one report for a different document. Students who learn (through lecture or reading) about audience analysis might be required to try out certain approaches: some that are likely to work for audiences close and well-known, others likely to be useful when the reader is more remote, and still others likely to serve when audiences are multiple and heterogeneous. Students can also be required to try out certain self-evaluation practices, to measure the efficacy of computer editing programs, to review documents collaboratively, or to test their documents on readers either informally or formally (see Goswami et al., ch. 8, 9)—all these methods empower revision. Some composing practice can be divorced from the production of documents: just as a basketball coach's players perform certain "unrealistic" drills to become more proficient once the games begin, so a writing coach can force writers to practice particular invention, arrangement, or revision maneuvers out of context in order that those manuevers will be available when a "real" writing situation arises. Of course basketball coaches do lots of scrimmaging, too, and so it also makes sense for writing instructors to ask students to produce lots of real or quasi-real documents.

When assigning certain documents, however, teachers would be wise to vary the composing situations: there might be in-class projects, short-term out-of-class tasks, and long-term assignments; there might be collaborative-writing assignments and individual ones; some assignments might be more important or more familiar than others; and so forth. Anne Herrington has pointed out that a writing class might be seen as a reasonable facsimile of a discourse community, complete with shared conventions, expectations, and constraints. Instructors who are aware of those conventions and constraints may use them constructively to help students grow as writers. For whatever assignments a teacher chooses—and my aim has been to suggest only a few possibilities— writers grow in direct proportion to the amount they are stretched.

And whatever the assignments might be, they will have to be evaluated in terms of how efficiently they are produced as well as in terms of how effectively they work. At the least, instructors must be as interested in commenting on processes as on written products. Instructors can review projects in the course of their development, overseeing plans and drafts and requiring students to reflect on processes before, during, and after their use. Once the project is complete, teachers can again force students to reflect as much on writing processes as on the products, perhaps by looking at the logs I mentioned earlier or perhaps by reviewing notes and rough drafts—and all the time suggesting alternatives or asking students to consider what they might do differently next time. A curriculum in the rhetoric employed in scientific and technical communities will be committed to making students flexible rhetoricians, capable of marshaling various options effectively in the service of various documents. But it will be just as committed to teaching students the flexible composing options necessary for realizing those various documents under various circumstances.

I know that a curriculum so demanding might well require more than one course or one year for prospective technical writers to negotiate. In that sense, what I call for may seem more ideal than practical, especially when many students are lucky to get even one course in technical writing. Yet it is also practical to remember that only when students have a rigorous program in composing theory, substantial practice, and careful evaluation will they become adept at composing in different circumstances at work and will they equal the best stylists in meeting the needs of different audiences, subjects, aims, genres, and settings.

Technical writing can profitably be seen as a contribution to an ongoing conversation taking place in a technical or scientific setting. As such, as James Reither has said, "what writers do during writing cannot be artificially separated from the socio-rhetorical situation in which writing gets done" (621). That fact has to be remembered by researchers into the writing process and by those technical writing teachers who wish to apply that research to the classroom.

PART THREE:
ON THE PROCESS AND PRODUCT
OF TECHNICAL WRITING:
CONTEMPORARY PERSPECTIVES

THIS SECTION BEGINS with the process of audience analysis and ends with the product evaluation, which is then incorporated into the process of revision. If the recurring theme in each essay is the critical relationship between the reader and the writer, the spirit of each essay is one of questioning venerable precepts and current practices in order to improve both the process and the product of technical writing.

As a crucial first step in creating technical documents that are easy to understand, audience analysis affects every aspect of the writing process: choices of content, arrangement, words, syntax, and visual display. Yet our traditional matrix for analyzing audiences for technical documents may be inadequate, according to Jo Allen. We must go beyond the static methods of analyzing levels of audience and examine reader apprehension, expectations, and protocols.

The new audience for technical documents is changing as well. With electronic equipment available to almost everyone, the audience for technical documents is rapidly becoming more widespread and diverse yet less knowledgeable about the subject of technical documents—increasingly sophisticated, complex hardware and software. Because the prevalent mode of technical documentation is descriptive when it should be instructional, Janice C. Redish and David A. Schell argue that technical writers must transform "content-oriented descriptions into user-oriented instructions." Further, technical writers should evaluate the effectiveness of the instructions they write, employing such methods as user-edits, protocols, beta trials, and laboratory-based testing. These methods not only pinpoint where a document needs to be revised but also identify stylistic features that aid or impede reader comprehension. Thus, document testing becomes part of the revision process and, at the same time, adds valuable content about the connections between style and readability to our research base.

To accommodate advanced technology to the new, inexperienced audience, the technical writer must be skilled in nontextual layout and design principles, or the arrangement of text and graphics as it affects the reader. Mary Lay posits that the technical writer must also become a technical designer with mastery of the principles of spatial unity, balance, proportion, emphasis, and sequence.

Yet technical writers must not be so swept away by the marvels of electronics that they neglect stylistics and, more specifically, the role of cohesion in a reader's understanding. The last two essays in this section study two long-cherished aspects of cohesion: maintaining parallelism and avoiding vague pronoun reference. Both essayists begin with traditional textbook rules and compare these rules with what experienced writers do in good technical and scientific prose. While Isabelle Kramer Thompson's study reaffirms the rule that parallelism increases coherence and clarity,

Michael P. Jordan's negates the rule that vague pronoun reference always decreases coherence and clarity.

Individually and collectively, the five essays in this section challenge a number of traditional precepts of our discipline.

Breaking with a Tradition: New Directions in Audience Analysis

Jo Allen

Since the 1960s, technical writing has been influenced by several landmark works from top professionals in the field. Two works, Kenneth Houp and Thomas E. Pearsall's *Reporting Technical Information* and Pearsall's *Audience Analysis for Technical Writing*, are especially well-known because they brought the concept of audience analysis to the forefront for writers. While it would be unfair—and inaccurate—to claim that no one considered audience before the 1960s, these scholars developed the first system for evaluating an audience. Based on their works, writers, teachers, and researchers have a method—composed of a series of questions about the reader's background, education, position in a company (i.e., use for the writing)—that guides writers through various options to make their writing appropriate for their reader.

For the past ten to twelve years, however, an unsettled tone has been creeping into literature and discussions, challenging these traditional concepts of audience analysis. Researchers, writers, and teachers have begun to recognize and address their dissatisfaction with the static methods of analyzing audiences. According to current research, the problem is that many of us (whether writers, researchers, or educators) have allowed traditional audience analysis to dominate our ways of thinking about readers. We have come to insist that only demographic factors, such as background, age, occupation, and familiarity with subject matter, can affect our understanding of audience. We have, in fact, stagnated a process—and the definition of "process" should tell us that we are doing something wrong. We have violated the intentions of the original authors, who suggested these factors should guide our thinking about readers but not control it.

Furthermore, other researchers, beginning with Walter J. Ong and followed by Douglas Park and Russell Long, are convinced that the very notion of analyzing an audience is misguided and outside the realm of what writers do. Central to this dissatisfaction is the controversy over whether writers should, in fact, analyze and then address an audience (the traditional concept of audience analysis) or whether they should construct roles that flexible readers may choose to adopt. The solution to this controversy will likely hinge on the writer's motivation to accommodate the audience. For example, the professional technical writer who writes safety manuals that technicians must understand is in an economically and occupationally precarious position that demands knowing the audience and then constructing the appropriate text. Other writers—even technical writers—who do not have so precise an audience or so perilous a mission may, in fact, relax into a more egocentric role as a casting director who may tell readers, in effect, that they are or are not suited for the part.

In addition to showing their dissatisfaction with traditional audience-analysis techniques, researchers are also questioning current ways of thinking about the writer-reader relationship. These researchers, approaching audience analysis from the perspectives of rhetoric, composition, and scientific or technical writing, have produced new methods for defining and studying audience analysis (how it works and what it encompasses).

Of primary importance is what we use as a basis for discussing audience analysis, and two works have emerged as essential for defining audience analysis. First, "Audience Addressed/Audience Invoked: The Role of Audience in Composition and Theory," by Lisa Ede and Andrea Lunsford, describes the controversy between the ideas of Ruth Mitchell and Mary Taylor and those of Russell Long and Walter Ong. It asserts that Mitchell and Taylor falter by giving the audience too much power while Long and Ong falter by giving the writer too much power. It posits that the best stance toward audience is recognizing that an audience is in some part invoked (created) and then addressed. Combining the insights of Mitchell and Taylor with those of Long and Ong, Ede and Lunsford create a homogenous model—one they hope reflects "the intricate relationship of writer and audience to all elements in the rhetorical situation" (169).

For the technical writer, this kind of homogenous model means recognizing that readers are not always "required" to read particular documents; sometimes they choose what they want to read. Motivation may make significant differences in the way readers read: the amount of time they are willing to invest in reading a document, their attitude toward it, the kinds of information they are most likely to garner from it, their expectations of it, and other factors that would require the technical writer to go far beyond the concerns of traditional audience analysis.

Adopting and expanding on Ede and Lunsford's work, Robert G. Roth finds that good writing is often a process that totally excludes considerations of audience or that allows the writer to change—substitute, narrow, or broaden—the intended audience. And, if the audience may indeed change during the writing process, then Roth's idea of the evolving audience subscribes to the "audience-invoked" model. He concludes that if writers do acknowledge that the audience may evolve during the writing process, then we can retire the question "Who are my readers and what do they need?" in favor of "one that encompasses the rhetorical situation in its entirety: 'On what basis do I claim my readers' attention?'" (55). Again, such an approach is situation-specific—with more appropriate implications, perhaps, for someone writing a sales or persuasive piece (which must entice the reader to read) than for the technical writer creating a proposal, report, or manual (whose information is often compulsory for the reader).

The second work that helps define audience analysis is Barry Kroll's "Writing for Readers: Three Perspectives on Audience." Kroll suspects that audience analysis means different things to different people. Calling the three perspectives on audience analysis "rhetorical," "informational," and "social," Kroll defines his terms, examines some of the underlying theoretical assumptions about each, illustrates their implications, and proposes objections and limitations to each. Looking for the same balance Ede and Lunsford seek, Kroll concludes "how unwise and unproductive it would be to swing from neglect of the audience to overemphasis, forgetting in our new enthusiasm the old lesson that writing involves a delicate balance among several elements of communication, of which audience is but one" (183).

With these cautions to remember the necessary equilibrium between the audience and other components of the writing process, we can move more easily into current research on audience analysis.

Scope of This Essay

Several problems in researching audience analysis prevent a comprehensive review of the literature. The primary difficulty is that audience analysis affects everything about writing: style, format, layout and design, documentation, graphics, vocabulary, definitions, analyses, descriptions, letters, memos, reports, brochures, manuals, and proposals. Discussing the role of audience analysis in each of these applications would require a separate, weighty anthology.

Aside from the problem of discussing audience analysis as it relates to technical and scientific writing—application, theory, research, and pedagogy —is the problem caused by the interdisciplinary approaches to the topic. Audience analysis does not descend from the writing disciplines but rather extends from—and into—the fields of psycholinguistics, cognitive psychology, philosophy, speech communication, reading theory, and rhetoric (which may be applied to various other fields that are at least tangentially related to theories of scientific and technical writing).

To combat these problems, I have limited the primary emphasis of my essay to works that focus on audience analysis in technical writing or whose implications have special relevance for technical writers. Further, the essay is limited to works on audience analysis published since 1982. (For a review of the literature before this period, see Michael Keene and Marilyn Barnes-Ostrander's "Audience Analysis and Adaptation"; for a history of the developments in audience analysis concepts published in *IEEE Transactions on Professional Communications*, see Robert Fry.) Thus, the essay is designed to provide cogent and insightful developments that may produce long-lasting effects on ideas about audience analysis in technical writing.

The essay first addresses research within the writing field itself, then research from other disciplines. Following this review of research are classroom applications of the research. Finally, other questions about audience analysis —its methods and implications—remind us that the research, like audience analysis itself, must be an ongoing process.

Current Research on Audience Analysis from Writing

More apparent in current than in previous research on audience analysis is the interest in the link between reading and writing. Going beyond the general assumption that good writing and good reading habits are related, current research explores the cognitive processes, attitudes, and preferences constituting these two endeavors and may come closer to solidifying a reason these abilities are related. (See, e.g., Salvatori; Silvers; Belanger and Martin.)

One of the most dominant—and hotly debated—topics in research is protocol analysis, which holds great implications for studying the writing-reading link, if researchers can resolve objections. Protocol analysis requires a subject to write (or read, depending on the process being evaluated) and speak into a tape recorder, registering the cognitive processes going on while the subject composes (or reads). Those who believe protocols have no value base their opinions on some of the observations David Dobrin makes:

1. Proponents of protocol analysis have not described their model—a necessary component of empirical research (713);
2. While protocols could possibly work to investigate problem-solving techniques (such as computers use), all writing is not suited to a problem-solving approach (716);
3. Protocols are distorted because subjects know what is expected of them (718); and
4. Protocols do not account for normal distractions and daydreaming or mind wandering ("Protocols" 722–23).

There are more objections, but there are also replies, such as those synthesized in Erwin Steinberg's essay:

1. Opponents have confused the terminology, failing to distinguish guided protocols from stream of consciousness (697);
2. "Protocols reflect cognitive processes adequately enough to build representative problem-solving models from them" (699);
3. Anything being observed may be distorted—a fact inherent in the scientific method (699–700); and
4. Protocols, which require immediate reactions, are more reliable than

retrospective analyses, making protocol analysis the best method for analyzing cognitive processes as they occur (708).

The value of protocol analysis is that writers can get immediate feedback from their readers, seeing what is unclear, ambiguous, or inaccurate about their message. For the technical writer, such an approach may, perhaps, become a viable test for the readability and usability of technical documents. The controversy continues, but protocol analysis remains a possible technique for analyzing the process of writing and the reader's reaction to that writing. (See Janice Redish and David Schell's essay in this anthology.)

Seth Finn provides another direction for studying the writing-reading link in audience analysis. He categorizes three levels of reading as these levels affect reader enjoyment: the decoding level (semantic-syntactic relations), the entertainment level, and the informational level. Although technical writers are not generally concerned with how much a reader enjoys reading a technical document, we have yet to study what effect reading pleasure has on attention, processing, and recall.

Finn finds that on the decoding level, readers enjoy unusual or unexpected vocabulary, provided the words are arranged in a predictable sequence (361). On the entertainment level, readers prefer articles that are characterized by novel and unpredictable content—up to a point. According to Finn, "In contrast to traditional approaches to readability, the model assumes that at this level, unpredictable content provides a pleasurable stimulative effect that compensates for the additional mental effort it demands from the reader" (361). Whether readers enjoy a composition at the information-seeking level—the level of greatest concern to the technical writer—depends on their ability to integrate the information into their present cognitive constructs (storage areas, so to speak, for information they are already familiar with). Finn explains:

> [I]nformation that conflicts with the reader's attitudes, beliefs, opinions, or any other expectation of reality can increase uncertainty, thereby affecting anxiety . . . that diminishes reading enjoyment. By contrast, positive effect most likely results only when information can be comfortably integrated into existing cognitive structures, thereby adding to the reader's store of personal knowledge. (361)

Other research addresses mutual knowledge—the concept that readers and writers share large amounts of knowledge about each other. The concept is based on an understanding of intentions and presumptions, elaborated by Gordon Thomas as

1. Knowledge of conventions—knowledge of regularities in language and format (about, for instance, grammar, punctuation, and the physical form of, say, a memo) (586);
2. Knowledge of language—the audience recognizes the writer's intentions

and the writer expects that recognition. In other words, there are no secrets between the writer and the audience (587); and

3. World knowledge—the audience has some presumptions about what the writer could say. Based on that expectation, the writer says something new because writer and reader share some basic, though implicit, knowledge about the world—"facts, common opinions, and so forth" (587).

Though novice technical writers may be dangerously susceptible to ego-centric expectations or assumptions about their reader's knowledge, Thomas finds that "true engagement occurs when the writer uses Mutual Knowledge to alter the beliefs, attitudes, and knowledge of a plausible audience" (593). By initiating discussions within the confines of an audience's knowledge and expectations, technical writers can impart new knowledge. The analysis of an audience, therefore, depends, not on demographic or even situational analyses, but on the technical writer's understanding of what the audience can be expected to know, enabling the writer to determine where to add detail, descriptions, definitions, analogies, or other aids to understanding.

Current Research on Audience Analysis from Other Disciplines

As interesting as the research within our field is (and I have touched on only a few samples here), other fields are steadily producing their own research, some examples of which are noted briefly as avenues for interdisciplinary research on audience analysis.

From the fields of psycholinguistics, cognitive psychology, and reading theory has come research that often parallels the kinds of distinctions we in writing are trying to make. Closely akin to the theory of mutual knowledge and given-new cognitive theories is the concept of prior knowledge (again, the types of information readers held before—and, thus, bring to—reading). In fields outside writing, researchers are studying how prior knowledge affects the reader's comprehension and how writers can use the concept of prior knowledge to enhance their readers' chances to understand material more easily. (See, e.g., Castleberry; Lipson; Baldwin et al.; Mosenthal et al.)

A related study by Ian Begg et al. researches the correlation between memory and believability, concluding that biases and familiarity with a subject affect the interpretation of new information. Further, they find that subjects are more willing to believe new information about previously familiar subjects than "new" information about unknown subjects. And since users of technical documents generally read for new information, perhaps the addition of comparisons would help them adjust more easily.

Researchers from various fields often focus on how readers read; their work holds great promise for learning more about how technical writers should write. Over the years, for instance, writers have looked for a mathematical measure of readability, which has produced—and extinguished—numerous formulas for assessing whether the intended audience can comprehend a document. While some work with formulas continues (see, e.g., Griesinger and Klene; Chandani; McClure; Ng), other research is concentrating on specific causes for improved or weakened comprehension. Looking beyond syntactic and semantic manipulations, one area of study concerns the effect visual display (layout and design) has on readers. John Morton et al., for example, have explored the complex relation between recorded (stored) information and cues for recall, concluding that people are more likely to understand and remember information that is signified by a code—perhaps the technical writing equivalent of a heading (20). Sandra Moriarty and Edward Scheiner have studied the effects of close-set text type and have found that readers read more words when the text is closely set than when it is traditionally spaced (702). Paul Kolers concludes from his study that readers use different processes for reading the same information presented in different typefaces, processes that are "remarkably sensitive to minor variations in type" (238). (See also Cocklin et al.; Underwood and McConkie; Blanchard.) When designing technical documents—especially user manuals—technical writers and document designers should consider these variations in analyzing their audiences.

Other researchers are more concerned with reading or comprehension. Particularly interesting are the studies proposing "that what is remembered over longer time periods is the meaning of the material read, divorced from the specific means by which it was acquired" (Kolers 232). (See also, e.g., Seifert et al.; Anderson, *Architecture*; Just and Carpenter.) If readers do incorporate their own meanings of text—often by pausing to construct summaries or clarifications—then they are, in effect, replacing the actual text with their summaries. What does that mean for the conscientious technical writer attempting to meet the readers' needs? Obviously, these writers may want to build in summaries at regular intervals to help direct those readers who will only remember the summaries.

Even more complicated research about the composition and functions of the brain and its abilities to process information is continuing. Endel Tulving has published several works since 1972 espousing the theory that there may be two memory systems: the semantic system and the episodic system. Calling these systems "parallel and partially overlapping," he distinguishes between the two: "Episodic memory . . . deals with unique, concrete, personal, temporally dated events that the rememberer has witnessed, whereas semantic memory involves general, abstract, timeless knowledge that a person shares with others" (307). The theory has generated as much debate in cognitive research as protocol analysis has in the writing fields. (See, e.g., McKoon et al.; Anderson and Ross; Ratcliff and McKoon.) But, again, if the conflicts can be resolved to support the essence of the theory, we in technical writing stand to learn much about the writing-reading link and to move toward resolving the discrepancies between theories of audience analyzed and accommodated (by writer) and writing analyzed and adopted (by reader). Further, we would have to consider whether the process of writing from episodic memory differs from the process of writing from semantic memory. How is the technical writer—who often writes from secondhand experience—affected? How would two memory systems affect theories of prior knowledge? And could semantic memory construct the background material on which to base new or episodic details?

Pedagogical Implications and Applications

Evidence of the dissatisfaction with traditional audience analysis is apparent not only from the theoretical research but also from pedagogical research and applications. Several researchers and educators have tried new methods for teaching and directing students' learning about audience analysis.

Applying the concept of protocol analysis, for instance, David D. Roberts and Patricia A. Sullivan state that protocols are useful "to achieve an understanding of the reader's responses (e.g., confusion, boredom, enthusiasm, misunderstanding) and an articulation of the causes of those responses" (148). They go on to express the benefits of using protocols: (1) Students may begin to extend their awareness of audience beyond prewriting and into the actual composing of the text, and (2) students may come to understand better the way readers read (152).

In "Analyzing Audiences," Douglas Park takes a different approach to restructuring pedagogical techniques for audience analysis. For students writing essays, he asks, "What must they think about to imagine their papers as having or capable of having an audience?" (479). Adopting a common technical writing pedagogical technique, Park uses situational analyses to give students an appropriate context for their writing assignments. But beyond the application of this technical writing strategy in composition classes, he suggests that teachers should (1) learn more about "defining the social functions of various kinds of public discourse," (2) "learn more about how different kinds of discourse written for public or 'general' audiences actually work rhetorically," and (3) recognize that "the culture of the classroom can be a persuasive influence on a student's ability to understand an audience" (487–88).

These suggestions should be taken back into the technical writing classroom, where situational analyses are most readily incorporated as a way of imitating the kind of writing students will have to do in their careers. Teachers of technical writing must recognize the limitations of such an approach: though this kind of in-class practice may promote the understanding that audiences do vary, students will meet quite different audiences outside the classroom.

R. J. Willey et al. offer yet another strategy for teaching audience analysis. Building on Donald Rubin's concept of "episodic perspective taking," these authors recommend that students learn to focus on "the readers' knowledge, feelings, and opinions with respect to the topic at hand, and with readers' perceptions of the social meaning of the particular written interaction in which they are engaged" (Rubin 234).

First, they cite Jeff Schiff's strategy, published in the *Arizona English Bulletin*. Schiff recommends that teachers and students accept the teacher's role as audience. He has students draw up a list of thirty questions to find out as much as they can about their reader (the teacher). The relevance of students' questions to analyze their audience guides them "into taking an episodic perspective" (Willey et al. 8).

Next, the authors cite Fred Pfister and Joanne Petrick's narrower but similar approach: directing students to question their audience's (instructor's, classmate's, magazine reader's, or others') feelings and knowledge about a specific topic, eliciting answers that guide students in deciding how much their readers know or need to know about the topic.

Finally, Willey et al. recommend another variation on episodic perspective taking: peer-group workshops and conferencing. They say:

> Group work provides students with the most fruitful experience in audience awareness, and when instructors involve themselves intimately in the work of peer groups, students can learn to take the perspective of the teacher-as-reader, often by discovering how similar the needs of peer-audience and teacher-audience are. (9)

According to the authors, students often begin passively but end up actively participating in the group, "asking questions, posing possible alternatives, probing and taking the perspective of [their] audience" (10). In the light of earlier criticisms that audience analysis is stagnant as a prewriting activity, and presupposing Roth's later discussions of the evolving audience, Willey et al. add that their research shows peer and instructor reviews to be "most valuable during the writing process and that taking an episodic perspective is more valuable at the revising stage than at the prewriting stage" (9). Recognizing that students in the technical writing class will have widely different audiences once they begin their careers, the technical writing teacher can use peer workshops to help students prepare for the audience variations that may affect the kind or style of writing they produce.

What remains is the unending list of questions we must try to answer about the crucial relationship between the writer and the reader. Lisa Ede has insightfully posed many questions in her article on audience analysis (144–45). In addition, we may also want to consider some of the following questions:

1. Does writing *to* an audience cripple the reader's growth in sophistication—cause the reader, in other words, to stagnate at one level of reading?
2. In what way does audience analysis skew the tone of writing, especially persuasive writing? Or, if readers do, in fact, "try on" reading as they do clothes, in what way does tone affect their decision to stop or continue reading?

3. What is the best way to determine the line between the writer's assessment of the audience's knowledge and the writer's own egocentricity?

4. What is the process of recognizing an audience as hostile, and how does knowing that an audience is hostile affect writing to that audience? How concerned with hostility can the writer be who believes audiences are created?

5. What effects do subtle geographical changes make in writing for an audience (between one rural community and another, for instance)?

6. What effect does a more radical geographic change make (between a major city and a rural community)?

7. What effects do significant social differences create?

8. How can we account for and address cultural differences in audiences— writing, say, for international trade negotiations?

9. In editing and revising, how does a writer or an editor determine what can be fixed for a particular audience and what demands a start-over?

There are many other questions to be sure, but these indicate that we are, indeed, changing our ideas about audience analysis. Breaking the mold of how we have traditionally treated any topic is a complicated process, perhaps best recognized by—or at least best described by—Thomas Kuhn in *The Structure of Scientific Revolutions*. While what is happening with research and ideas about audience analysis hardly compares with the scale of the paradigm shifts Kuhn addresses, on a smaller scale we are going through some of the processes needed for a paradigmatic shift. And our place seems to be at the beginning of that shift: we sense the problem with traditional audience analysis, but have found no model, no paradigm, to shift to.

Primary issues in current research present not so much a new model (though some theorists have tried—probably prematurely) as elements that should be included in a new model. At the beginning of our paradigmatic shift, the greatest concern should be that we do not limit our ideas or our questions about audience analysis. Stopping short of any real progress with any real answers oversimplifies the concept of audience analysis, a concept that certainly demands as much attention as any other part of the writing process.

Writing and Testing Instructions for Usability

JANICE C. REDISH AND DAVID A. SCHELL

AS JAMES SOUTHER describes earlier in this book, in the past decade technical writing courses and degrees have proliferated on both the undergraduate and graduate levels. Many of the new courses and many of the new students are focusing on how to write manuals that tell people how to do tasks such as running a computer, operating a VCR, or filling out an insurance form.

In this paper, we explore two related questions:

Why is this new focus on instructions valuable for technical writing?
How can technical writers find out if their instructions are effective?

Focusing on Instructions

Much of the recent research on technical writing has focused on instructions for at least three reasons:

The field of technical writing has expanded greatly in the last ten years, and the new work is in writing instructions.
Technical writers have had to learn how to address larger and less technically sophisticated audiences.
Research has shown that people use documents as instructions even when those documents were written as descriptions.

Most new technical writing jobs are in writing instructions

Membership in the Society for Technical Communication (STC) more than doubled in the first half of the 1980s, and most of the new members are technical translators, writers who can turn descriptions of computer syntax into easy-to-understand instructions ("Technical Fields").

Many graduates of technical programs now find themselves writing instructions more often than they write descriptions or reports. In a recent survey, Paul Anderson of Miami University of Ohio asked graduates of seven of Miami's technical and scientific majors how often they write each of eleven types of documents. (Anderson's "What Survey Research Tells" is a comprehensive review of the results of approximately 50 surveys of writing in the workplace. For information about conducting survey research, see Anderson's "Survey Methodology.") For Anderson's 841 respondents, only memorandums and letters rank higher than instructions.

1. memoranda
2. letters
3. step-by-step instructions
4. general instructions
5. filling out forms
6. proposals
7. formal reports
8. minutes of meetings or conversations
9. speeches or presentations
10. advertising
11. articles for professional journals ("What Survey Research" 23)

If the goal of education in technical writing is to prepare students for the marketplace, classes in how to write clear instructions must have a prominent place in the curriculum.

Writers have to reach broader and less technically knowledgeable audiences

The mass market for electronic equipment has been the driving force behind the expansion of technical writing. Almost everyone must now deal with technical instructions—to use the computer on one's office desk, to use the VCR at home, even to find a book in the computerized catalog at the local library.

Therefore, technical writers must learn to communicate with new audiences, audiences that are larger, more diverse, and often less technically sophisticated than the traditional audiences for technical or scientific writing. Where technically trained readers may have a schema for reading technical material, the general public does not. Technical experts are often poor at explaining systems to novice users, because they overestimate the users' knowledge.

In addition, there has been a spillover from the concern for clear instructions for the mass market to a similar concern for technically trained users. As equipment becomes more complex, the amount of information needed to use and maintain it grows (Duffy). Even a trained specialist cannot be expected to remember everything, and the potential danger in not using or maintaining the sophisticated equipment properly may be much greater than with simpler, older technology.

Technical writers, therefore, must also find ways to make the necessary information immediately accessible and instantly understandable to the technical specialist. In the course of many projects, we have found that:

Technical writers tend to overestimate their readers' technical knowledge.

Even when they have a schema for extracting information from technical documents, technically trained professionals find it easier to read and understand instructions that address them directly.

Manuals are tools for busy people who are trying to get a job done; readers (including technically trained professionals) want to get to the right place in the manuals as quickly as possible. They do not want lengthy technical discussions; they want procedures.

Many documents that are not written as instructions should be

Readers who are looking for instructions (procedural information) too often find descriptions instead. Compare, for example, these two versions of one of the requirements for obtaining a student loan:

The descriptive approach: Disclosure of the applicant's SSN is required as a condition for participation in the Guaranteed Student Loan Program.

The procedural approach: You must write your Social Security Number on the application if you want a Guaranteed Student Loan.

Note that the first sentence is a statement of fact, but it does not tell the user what to do. One sentence like this is not too difficult for most readers, but imagine an entire document in the descriptive style. The facts pile up, and the reader gets lost.

A traditional computer manual is an example of page after page of descriptive facts from which users must infer procedures. When users go to a computer manual looking for instructions on how to accomplish a task, they often have to figure out the instructions for themselves from the commands or menu options. Because commands are usually described in alphabetical order, the reader has to decide first which command is likely to be the one to use. Then the reader has to translate the description into steps for action.

Here, for example, is a paragraph that is typical of traditional computer manuals.

The QUIT command terminates an editing session without executing a SAVEFILE. To terminate with an automatic save, the EXIT command should be issued. QUIT has no parameters and is accessed through the F3 programmable function key. Confirmation is required to complete execution of a QUIT; a negative response terminates execution and the cursor is returned to the previous location in the workspace file. When a QUIT is executed, the workspace is cleared and control is returned to the operating system.

What the user really wants are instructions like these:

Leaving the Editor

When you are ready to leave the Editor, you have to decide whether you want to save what you have just typed.

Leaving and saving your file
[describe how to "exit"]

Leaving without saving the latest changes to your file
If you do not want to save the changes you have made to your file since you last saved it, follow these steps:

1. Press F3.
   You see the message:
   Any changes that you made since you last saved your file will be lost. Continue quitting? (Y or N).

2. If you change your mind and want to go back to your file, type: N
   The Editor returns you to the file you were working on.

3. If you do want to leave without saving your changes, type: Y
   The Editor returns you to DOS. You see the prompt, A>. You can now turn off the computer or use a different program.

Note that the procedural approach does not mean only giving the steps. Users also need context for the steps (Charney and Reder). In our example, we begin with a task-oriented heading that ties the instructions to the task the reader wants to do (namely, leave the Editor). We then explain "when and why" before we explain "how" to use each method of leaving.

Although teachers of technical writing exhort students to pay attention to audience and purpose, we have found, in analyzing hundreds of technical and legal documents, that most traditional technical writing does not, in fact, address any audience directly or serve the purpose for which the audience uses the document (Charrow; Redish, "Language"; Redish, Battison, and Gold; Redish, "Writing").

In traditional documents (e.g., computer manuals, administrative manuals, insurance policies, employee benefits handbooks, and loan applications), we find that:

The information is presented in alphabetical order, or the order in which the parts of the product were constructed, or the order in which the information occurs to the writer.
The writing is noun-based.
Headings are single nouns or noun strings.
There are no pronouns.
The subjects of the sentences are inanimate.
The verbs are in the passive voice.

The writing describes a system, while users come to writing expecting instructions on how to act.

The people who have to read these documents find them difficult to understand. In a series of studies of a traditional, descriptive federal regulation, researchers from Carnegie Mellon University found that readers could not match the noun-based headings with the text but could match the headings and text when the headings began with relevant action verbs (Swarts, Flower, and Hayes). They also found that when readers were asked to talk aloud as they tried to understand the regulation, the readers translated the abstract, nominal writing into "scenarios"—verb-based sentences in which actors performed actions (Flower, Hayes, and Swarts).

Thus, the significant trend of the past decade has been to make technical documents more understandable to users. Most of our current work at the American Institutes for Research is to transform content-oriented descriptions into user-oriented instructions.

Studying Instructions with Usability Testing

When technical writers pay more attention to how they are writing instructions, they also become more interested in knowing if the instructions that they write are effective. The best way to evaluate instructions is to observe representative users trying them out (Atlas; Soderston, "Usability"). This type of observation, when it occurs in a controlled setting with researchers as observers, is called "usability testing" (Mills and Dye; Schell, "Usability"). Usability testing is often used to study computer manuals, but the techniques can be applied to any type of instructions. Although computers are often used as part of the technology—for example to keep a running record of the user's actions—they are not necessary. Paper and pencil work also.

What are the critical features of usability testing?

For a usability test to be informative, it must meet two criteria:

The test subjects must be people who might actually use the product.
The work in the test must be similar to the work that users would actually do with the product.

The techniques used to gather information during the test can vary. For example, the user may or may not be given instructions to talk aloud. The researchers may or may not videotape the test.

How does usability testing help?

Usability testing can be either evaluation or research. When we conduct a usability test as evaluation, we focus on a specific product and its documentation. We want to know if people can use the product and the manuals. Our goals may be to **verify** that the manual is accurate, to **validate** that people can use the manual to accomplish certain tasks, or to **diagnose** specific problems in the product and the manual—and to suggest solutions.

As we observe many users in many situations with many products and manuals, we can also use the results of our evaluations as a data base for research. When we conduct a usability test as a research project, our goals may be to **understand** how people typically use different types of manuals and to **compare** the effectiveness of a particular feature or set of features on how easily people can use a product or a manual.

What types of testing can one do?

Different types of usability testing serve different purposes. Each has advantages and disadvantages.

USER EDITS

In a user edit, the user is told to read a manual aloud, one page at a time, and to follow the instructions as closely as possible (Atlas). Thus, a user edit is a good way to verify that all the instructions are accurate and in the correct order. If users are able to follow the instructions page-by-page and step-by-step, then they should perform each task successfully. Furthermore, by noting where the user stumbles, hesitates, or misreads, the researcher can tell where the instructions need to be clarified.

A user edit is most appropriate for a short set of instructions that the user must follow from beginning to end—for example, the instructions for putting together a bicycle. A user edit is also useful in the early stages of writing a computer manual. Having even one user read through the instructions in the style that a team of writers is using can help the team see if it has developed an understandable style.

A user edit is not sufficient, however, as a usability test of a computer manual. Because the user works through the manual page-by-page, a user edit will not tell the writer if the user can find information in the manual (Atlas). But if users cannot find the information they need in the time they are willing to spend looking for it, they will not use the manual (Redish, Battison, and Gold). Moreover, most people do not read through a manual page-by-page (Atlas); instead, they try to locate the specific information they need (Sticht et al.). Even when they find the correct page, they do not willingly read large blocks of text (Carroll and Mack; Sullivan and Flower).

PROTOCOL-AIDED REVISION

In a protocol-aided revision (e.g., Flower, Hayes, and Swarts; Sullivan and Flower), typical users are asked to perform tasks similar to those that customers will eventually do with the product. They are also asked to talk aloud as they perform the task. The users are given the instructions just as they are given access to the product, but they do not have to read the steps. The researchers are interested in how the subjects use both the instructions and the product. The emphasis is on what the person is doing or thinking when performing each step in the task (Ericsson and Simon).

Protocol-aided revision is more realistic than a user edit in that the users must find as well as understand the relevant information. When conducting a protocol-based study, the researcher hears how the users interpret the instructions while they are trying to use the instructions. The questions users ask themselves, the problems they articulate, and the misinterpretations they make are all useful information. Furthermore, when users talk aloud as they work, the researcher can gain insight into the heuristics they are using to do the tasks.

We must distinguish protocols as a technique for understanding how people use and interpret a particular document from protocols as a record of human information processing. At Carnegie Mellon University, researchers are using protocols both to help develop clear instructions and to help understand cognitive processes. Currently, a controversy is raging in the literature over the question of whether what we learn from protocols gives us insight into human cognition. (See, e.g., Ericsson and Simon; Nisbett and Wilson; Cooper and Holzman; Steinberg; Dobrin, "Protocols.") But the question about cognition is a separate issue; we do not have to take sides in the controversy about cognition to consider protocols an extremely useful tool in evaluating instructions.

Talking aloud, however, is difficult for some people. To get around this impediment, in some of our laboratory-based tests we have two people working together. Through their conversation, we capture much of what one captures in a protocol; conversation is a more natural mode for most people than talking aloud while working alone.

BETA TESTING

Beta testing is the term that the computer industry uses for field trials of a product and its manuals. Beta testing is similar to usability testing in that users try out the product. Beta testing has the advantage of realism. The "test subjects" use the product in their own offices for their own purposes.

Beta testing is not as useful as laboratory-based usability testing, because a beta test is not methodical, nor is it monitored. The people who are trying out the product are asked to report any problems they find and to respond to questionnaires about how useful the product and documentation are. Beta testing has several disadvantages:

1. The researchers hear about only those problems that people bother to report. Lots more may have gone wrong that is not put down on paper.

2. People are reporting problems after the fact, so the researchers get no insight into what actually happened.
3. The researchers do not know what has and has not been tested.

What characterizes laboratory-based usability testing?

Laboratory-based usability testing, of the sort that we conduct at the American Institutes for Research, differs from user edits, protocol-aided revision, and beta testing in these ways:

1. It is more informative than a user edit—at least, for instructions that users do not typically read from beginning to end—because it tests how easily users can find what they need, as well as how easily they understand the information when they find it.
2. It gathers more data than protocol-aided revision alone. The protocol is only one of several sources of information about what the user reads and how the user acts.
3. It is more methodical than beta testing. In a laboratory-based usability test, researchers know just what is being tested and how the user and the product are interacting.

Like all the other evaluation techniques, laboratory-based testing has disadvantages. Like user edits and protocol-aided revision, a lab-based test is artificial. We try to make the situation as realistic as possible by arranging the physical setting to represent the appropriate environment and by setting the tasks that the subjects are asked to do in a meaningful context. (That is, we give the subjects "scenarios," or scripts, in which they assume realistic roles, roles in which the tasks are important and interesting.)

Nevertheless, users may act differently when they know they are part of a test than when they are using a product and instructions for their own purposes in their offices or homes. For example, if we are testing a tutorial, we usually ask people to go through the tutorial—and they do. Observations both in the lab and in our offices, however, show that, if they are left on their own, many people will ignore the tutorial and try to learn by "playing" with the program.

As with user edits and protocol-aided revision, cost and time require that we limit a usability test to only a few users. Typically, we have five to fifteen test subjects. The results of the test are useful only if we have chosen the subjects well (they must accurately represent the real users) and if we have chosen the tasks well (they must accurately reflect what users will want to do with the product).

In laboratory-based testing, we compensate for the paucity of subjects by the richness of the data we collect. We use several techniques to gather information, and we do so for two reasons. First, there is so much happening in a test that we want a record of it to study more carefully at a slower pace. Second, each technique gives us a different type of information; we have more confidence in our findings if we have several types of data that point to the same result.

For example, we use both videotape and audiotape to record what occurs during the test. Sometimes, actions happen so fast that no one is sure what the test subject or the computer actually did. We can review the videotape in slow motion to pinpoint the source of a problem. On audiotape, we capture both the subjects' statements and the researchers' observations.

We use a data-logging program to keep a coded profile of what the user does throughout the test. The logging program records the amount of time that it takes the test subjects to read specific pages and screens and how often they move back and forth between two screens or between the book and the screen. Sometimes we keep a list of the keystrokes that users make. And we use questionnaires to gather subjects' opinions of the product and the manuals.

In conclusion, let us pull our two topics together. Technical writing is becoming an increasingly important field because more people are using more technology today than ever before. In a highly technological society, clear instructions are critical. Nuclear power plants are run by people who use and monitor computer programs. The entire financial records of a company are kept in a computer data base—and clerks, as well as data-processing professionals, use those records.

Research has shown us what is wrong with old-style (descriptive) manuals and how to write new-style (procedural) instructions; however, even the best writers miss steps, become too knowledgeable about a product, or fail to match the tasks of the people who will actually use that product. Testing the instructions (and the product) with representative users is the only way to know if the instructions are useful. Testing can be an informative (and humbling) experience. Even expert writers are surprised by what readers do.

Nonrhetorical Elements of Layout and Design

MARY M. LAY

"ONE OF THE MOST CRUCIAL skills required of a technical editor is the ability to choose and employ graphic devices wisely," states an associate information developer at IBM. These graphic devices include "non-textual, non-content, non-prose related variables" such as typography, layout, illustration, color, paper, and binding (Soderston, "Evaluative" 87). With the advances in electronic publishing and word processing, the technical writer or editor now may control page layout as well as publication format. Technical writers are well versed in format specifications, such as levels of headings, and in prose-related considerations, such as tables and charts; however, with the freedom to arrange text and graphics on each page—and even on each computer screen—comes the responsibility to assess the effect of that arrangement. Which typeface will be most legible for engineers who must scan a manual for important information? Should more than one topic and graphic appear on each documentation page for secretaries learning to use a word processor? How much material on a computer screen can be emphasized by blinking symbols without taxing the viewer? The principles of design—unity, balance, proportion, emphasis, and sequence—provide answers to these questions. Even a brief introduction to these principles should give writers confidence when choosing these "non-textual, non-content, non-prose related variables."

Writers have long recognized their responsibility to offer accessible information. "Nobody curls up with a good manual and reads it from cover to cover," quips a technical writer with Standard Oil Company. "The reader is more apt to scan a few pages of the manual, consult it from time to time, or use it to learn a procedure. It is the technical writer's job to make it as easy as possible for the reader to get information in this way" (Zoe Brown 27). But the psychological impact of well-packaged information goes beyond objective or functional effect. "People seem to feel they get value," states an IBM advisory information developer, "from information that has eye appeal and seems to be and may well be easy to use" (Thing 9). A favorable impression bears even more on technical documentation of the late twentieth century, as the writer must accommodate technology to the noncaptive, inexperienced audience. A manager in information development expresses this difference: "The point is that if you feed indifferent computer documentation to non-captive readers, they may reject it" (Henderson 5). Moreover, the writer who displays information on a computer screen rather than on a printed page must consider that the primary impression is visual. According to one display-screen designer, "The first effect of a panel on a user is pictorial. Users see the layout and the overall pictorial effect before they recognize that there are words" (Ridgway 19). As the audience of technical documents swells with noncaptive and inexperienced users and as the technical writer maintains greater control over layout, effective design decisions will influence an audience's impression and eventual use of documents.

The serial approach to publication, in which the writer passes the draft on to the editor who sends the revision to a designer who then prepares the pasteup for the printer, has been made obsolete by electronic publishing. Documents are now more "writer-centered" than "product-centered," according to Bernard Peuto, president of ViewTech, Inc., and "the results are therefore much more likely to reflect the writer's original version of the document" (11). Many format and layout decisions should be made by the writer before the writing begins, as these choices will help the audience "interpret, comprehend, and remember the material," states Susan Grimm, author of *How to Write Computer Manuals for Users* (60). Traditionally *format* applies to the arrangement of information that appears on every page or screen, while *layout* refers to the arrangement of text and graphics on a specific page or screen and reflects both the format specifications and the unique characteristics of that page or screen. Format decisions may cover margin width, heading placement, and typeface and size, while layout decisions cover use of white space to retain unity, balance of text and graphics, and ways to emphasize important topics. While art designers have long applied these design principles to layout, writers—who now function as designers—need to learn the techniques of unity, balance, proportion, and emphasis and sequence.[1]

Unity

While some designers might begin their thinking with other principles, such as proportion, the writer-designer should let unity dictate initial choices since unity integrates all parts of the communication, including text.

Definitions

Unity determines the primary appearance of the whole page or screen. The parts or elements on a page or screen, including text, headings, and graphics, must look as though they belong together. Unity contributes to the coherence and legibility of the page or screen. Since we naturally seek organization, we see the entire pattern formed by the elements before we notice any individual elements. The writer-designer can achieve unity by three design techniques: proximity, repetition, and continuation.

Proximity involves empty space, sometimes called negative or white space, that can either unify or "blow apart" the elements on a page. Generally a writer-designer places the elements on the page close enough to each other so that more white space is external to the elements than internal, as in figure 1. If too much white space separates the elements, the design is chaotic and disconnected, as in figure 2; on the other hand, external white space can seem to press against the elements and unify them. To support proximity, the writer-designer can repeat certain elements or element characteristics, such as color or shape. Since similar things always appear to belong together, unity is achieved through repetition, as shown in the repetition of the rectangular shape in figure 3. Finally, a continuous line or direction that helps the audience move from one element to the other, such as the alignment of a list in figure 4, contributes to unity.

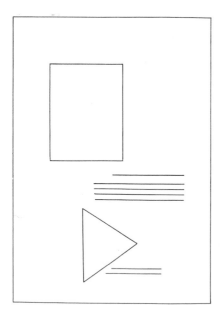

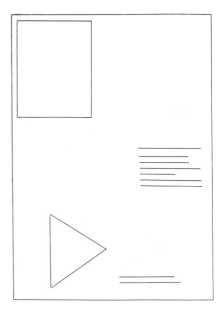

Fig. 1. Unity by Proximity. The use of external white space and the close proximity of the elements create unity.

Fig. 2. Lack of Unity. Chaos results from too much in-internal white space between elements.

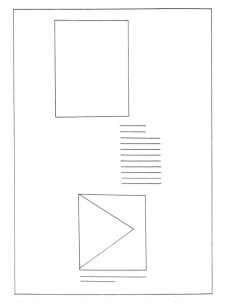

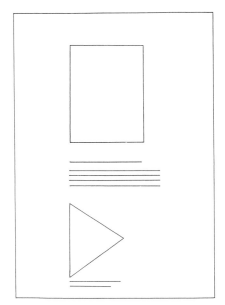

Fig. 3. Unity by Repetition. Repetition of the rectangular shape creates unity.

Fig. 4. Unity by Continuation. Aligning the elements on the left creates unity.

LIBRARY OF MOUNT ST. MARY'S COLLEGE EMMITSBURG, MD.

Applications

Since technical writer-designers now know audiences prefer that graphics be integrated with the text rather than gathered at the end of the document, text-graphic unity is essential (Pinelli et al.). The modular, as opposed to sequential, layout is the most recent result of this preference. In the modular layout, each topic or step in the document appears as a unit on a single page; the army's *New Look Technical Manuals* uses this type of arrangement. As figure 5 illustrates, each page is devoted to one step, and one step only. The step is explained in the text, which is accompanied by supporting graphics. Whenever possible, no text or illustration is carried over to the next page, and when the elements in a step occupy less than a whole page, the rest of the page is left empty. Even though the modular format increases page count, it helps the nontechnical audience complete an assigned task. The new-look manual is also unified by repetition; the same color highlights throughout, and the same generic human form appears whenever users need to "see" themselves performing the task. A writer and designer team, with overlapping skills and responsibility, produce the new-look manual for the visually oriented soldier and the task for which "technical competence is not a prerequisite" (Meyer 17). Although the basis for modular organization is function, the writer-designer must ensure that the text and graphics used to explain each task are visually unified.

Modular unity becomes even more important when the writer-designer displays material on a computer screen. As Dennis Charland from IBM comments, "A display terminal is more like a television set than a book. The information that is shown on the screen at any given time is like one picture or frame in a motion picture film or on a videodisc. The information module, therefore, should be designed as a sequence of frames that will be shown one after another on a display screen." In particular, viewers cannot easily refer back to one display after going on to another. To aid the audience, Charland feels that writers and designers should "think of the frame as one picture rather than as individual units of text and illustrations." Writers must learn to tailor what they say and the way they say it "to make each frame complete in itself" (160).

Frames should be consistently designed; the same element should appear in the same place for continuity and if possible be identified by borders or symbols unique to that element. The Poughkeepsie labs of IBM suggest in *Guidelines for Man/Display Interfaces* that these reserved areas be separated by white space or different lines such as dashed versus dotted lines or lines of various consistency, intensity, or color (Engel and Granda 8). While color in print and on screen contributes to unity, the writer-designer must recognize that not all colors are equally discernible and equally readable; in fact, color varies greatly from monitor to monitor and from system to system, so that the writer-designer has little control over what color the viewer actually sees (Murch 19; Rubens and Krull 32).

The final challenge in maintaining unity in display screens involves white space. Proximity supports unity; but liberal use of white space becomes even more essential on the display screen, even though the screen can hold fewer characters than the typical page and some writer-designers unfortunately want to pack the screen with information. The audience tires more quickly when dealing with this medium and, of course, cannot flip back easily to preceding material.[2] The writer-designer must remember that the extra white space should be external. Lines should be shorter and top and bottom margins larger; the large areas of white space should not be placed between elements.

Balance

Just as lack of design unity disturbs the audience, imbalance on a page or screen distracts. The next two design principles, balance and proportion, offer guidelines on how best to place the parts of the whole unified page or screen.

Definitions

On first sight, we imagine an axis on any page or screen and expect that what appears on one side of the vertical line of the axis will "weigh" as much as what appears on the other side. This vertical line serves as a fulcrum on a scale;

equal visual weight on either side gives the reader a sense of equilibrium. Large, unusual, or colored elements weigh more than small, ordinary, or black and white elements. A large area of white space also appears heavy. If elements on a page or screen are not in balance, we feel a need to rearrange.

While a writer-designer may balance an area symmetrically (mirror images on each side of the imaginary vertical line) or asymmetrically (different images on each side but with equal total weight), too often technical documents have been automatically balanced symmetrically. Repeating similar shapes in the same positions, such as a two-column format with centered headings (fig. 6), gives a formal but monotonous and imposing image. The audience begins to long for variety or to focus attention elsewhere. Interesting asymmetrical balance is achieved when dissimilar objects are balanced on the page; for example, in figure 7 the one-column format balances headings and illustrations beginning in a wide internal margin with a narrow external margin, heavy text block, and extra white space. The page still is balanced and stable but visually attractive; its appeal is especially important to the noncaptive audience whose attention must be caught and held.

Applications

Zoe Brown of Standard Oil Company believes that "findability" is the advantage of asymmetrical balance. To have findability, a layout should be consistent and predictable, with information blocked and labeled and with easy-to-scan internal heads. Spilling these heads over into a wide margin with the text off-center helps the audience scan for information (see fig. 7). The text lines should be shorter, and headings—"items of control information," as Brown calls them (28–29)—appear more frequently. While the symmetrical format accommodates more information per page and creates a formal appearance, the new asymmetrical formats scan more rapidly and hold the attention of the noncaptive, inexperienced audience.

The army's new-look manuals employ an asymmetrical balance in one-column text format. Large illustrations on one side of the page are balanced by white space on the other (fig. 5). Not filling that white space often preserves the modular approach. Similarly, in some IBM two-column formats, one column (often the inside column for left-hand pages and the outside column for right-hand pages) is either left blank or reserved for illustrations, as in figure 8. Graphic placement dominates the layout of each page. If the graphics that pertain to a topic take more room than the text, the text column is left blank until the writer-designer can again coordinate text with graphics. If the graphic is too large to be contained in one column, it spills over into the text column. Noticeable white space left whenever text runs out or when graphics dominate balances the page; empty space again can "weigh" as much as a large graphic. Also, the white space then adds balance and relief for the reader who appreciates a layout not filled with information. Finally, other asymmetrically balanced documents reflect what was originally called the "Midwest" style (fig. 9). Three

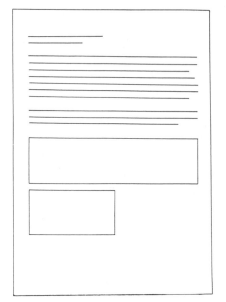

Fig. 5. The New-Look-Manual Layout. Only one topic or step is presented per page. The text explains the topic; supporting graphics follow.

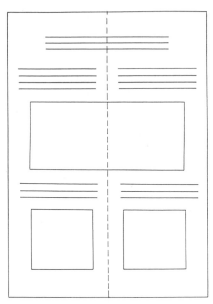

Fig. 6. Symmetrical Balance. The dotted line represents the imaginary line of an axis that divides the page in half. Right and left sides are mirror images.

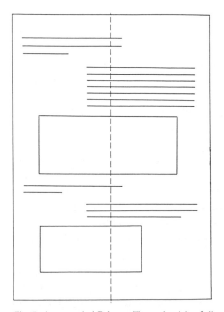

Fig. 7. Asymmetrical Balance. The total weight of all the elements on the right side of the imaginary line equals that of all the elements on the left side.

Fig. 8. Layout for Many IBM Manuals. Placement of graphics dictates page layout.

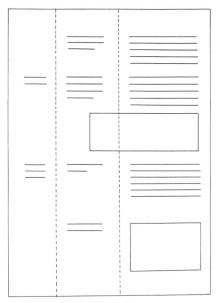

Fig. 9. Midwest Style Layout. The dotted lines represent the divisions between the unequal columns on a page.

columns of unequal width organize the page. One narrow column stays blank or contains headings (usually the inside column for right-hand pages and the outside column for left-hand pages), the next column gets wider and includes text and graphics, and the last column, the widest, provides even more information. This pattern introduces the audience gradually to more and more data but still enables the scanner to spot reference points easily.

Proportion

A writer-designer usually considers proportion and balance at the same moment. While the overall proportions of a page or screen or the overall proportions of each element—text, illustration, heading, and such—must also be considered, the ratios or proportions created when elements are placed in relation to one another have more effect on the audience.

Definitions

Since proportion refers to the relative size of elements and of the page or screen as a whole, after seeing a page or screen as a whole unit the audience then divides that page or screen into distinct visual parts. The audience at least compares the size of the page or screen devoted to text to the portion devoted to graphics. In comparing proportion the audience may concentrate on the horizontal line of the imaginary axis (in contrast to the imaginary vertical line that "balances" the page or screen). Where this horizontal line intercepts the vertical line depends on visual divisions; for example, the audience may see a line where the text ends and a graphic begins. Just as symmetrically balanced areas are formal and static, pages or screens where elements have equal proportions are uninteresting and confusing. A page or screen divided into a 50:50 ratio is monotonous, and it may give the audience no indication of which is the more important portion, as in figure 10. Designers prefer uneven proportions, such as a page that is one-third text and two-thirds graphics (fig. 11), just as they prefer an odd rather than even number of elements, such as three illustrations per page rather than two or four. Even if major elements do not appear to create visual divisions, the audience will still visualize the imaginary axis with the horizontal line well above the center of the page. Thus the optical center of the page will appear about one-third of the way down from the top of the page or screen. With no elements to guide us, we still look for uneven visual divisions. A general design rule is that meaning lies in contrast: the unusual, the irregular, or the large attracts our attention.

Applications

The technical writer-designer may not have much freedom in choosing the proportions of individual elements on any one page, because content may dictate proportion. The writer-designer may manipulate documentation format or size, however, to reflect interesting and appropriate proportions. Two technical writers at R. R. Donnelly & Sons note, "For example, when writing for technicians,

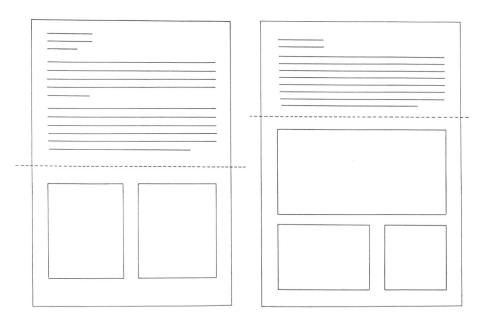

Fig. 10. **50:50 Proportion.** The dotted line represents the imaginary horizontal line that separates the main elements. The audience does not know where to focus.

Fig. 11. **Uneven Proportion for a Page.** The audience knows that the largest graphic is the most important and that the total graphic area is more important than the text. Using three graphics, rather than two or four, adds variety to the layout.

we use a two-column format, one for text descriptions and one for annotations for further reference. For operators, who need to step through a complicated procedure, we use a horizontal page format that enables the manual to be secured on its side in a flip-chart manner" (Estes and Rojecki 101). The size of the whole manual is changed to accommodate the second type of user; a saddle binding (stapled through the inner margin or gutter) or spiral binding rather than a perfect binding (spine-glued) ensures that the manual will stay open. Increasing the relative size of white space, margins, type, and illustrations helps the inexperienced audience handle complicated subject matter. Creating interesting page proportions is most challenging when using a one-column format, the preferred format according to Pinelli, Cordle, and McCullough's survey of technical audiences (77).

The square proportions of a computer screen dictate shorter line and paragraph lengths. According to Linda Chavarria of IBM, paragraphs in computer displays should be half those on the printed page and should be justified (aligned) on the left but not on the right-hand margin (ragged right). Chavarria prefers an "offset" style of uneven proportions, with names of fields or options listed in a narrow column on the left of the screen and wider paragraphs of information on the right, to give the reader visual rest and variety. Here proportion can be judged by both imaginary horizontal and vertical lines, as in figure 12. Finally, if writer-designers simply remember that the greater portion given to an element the greater the importance, they can begin to guide their audiences through the information.

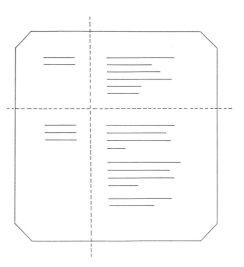

Fig. 12. **Uneven Proportions for a Screen.** Interesting, uneven proportions for a display screen, using the "offset" style. Dotted lines represent the points at which the audience will naturally separate elements.

Emphasis and Sequence

While writers and designers usually decide immediately what is the most important element on the page or screen, they must remember that the audience looks for visual cues about where to start.

Definitions

The final two principles of design are seldom separable, because the element of greatest importance or emphasis in a design initiates the sequence or direction of eye travel through the page or screen. Again, the largest element, the one in color, or perhaps the one with the most unusual shape attracts attention first. If one element is not outstanding, we begin to read or view in a "Z" pattern from the upper left hand corner of the page or screen. In the medium demanding greatest visual control—advertising—emphasis and sequence lead us from illustration and headline to copy, or in cigarette ads away from the required, but to the manufacturer undesirable, surgeon general's warning. In technical documents, emphasis and sequence, while more subtly indicated, are still essential tools for the writer-designer.

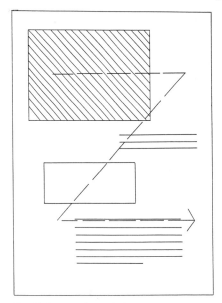

Sequence or eye travel depends on the illusion of motion or line. We tend to seek motion in a design; we consider horizontal lines static and diagonal ones dynamic. Repetition of shape indicates motion as well as unity. Actual lines—such as justified text or tabular rules—control sequence, as we move from line to line or from number to number. The implied lines in figure 13, however, also contribute to the natural "Z" sequence; just as we mentally connect a series of dots to form an image, we connect elements that are close to one another, such as the letters in a word or graphics in proximity. Just as we need unity, balance, and interesting proportions in layout, we need to know what to look at first, second, third, and so on.

Fig. 13. **Sequence by Implied Lines.** Sequence as controlled by implied lines as elements almost touch each other and as controlled by the natural "Z" pattern.

Applications

On the page, one of the most effective ways to emphasize is to display in color. If the writer-designer uses one color for emphasis throughout the document, the audience immediately knows what is of greatest importance on each page. For example, in the new-look manuals, solid green, red, orange, or brown indicates the point of interest. To preserve unity, shades of the same color depict background areas or subsystems.

On the computer screen, key elements are highlighted by increased intensity, use of reds or oranges, flashing or blinking elements, underlining, changes in type size or style, pointing arrows, reverse type in a light box, or borders surrounding the key element (Engel and Granda 3). The challenge in choosing among these methods is in balancing emphasis against eyestrain. For example, blinking, very effective for urgent messages, taxes the viewer the most. Some writers and designers have a definite preference as to where the most important elements should be placed; Rubens identifies the upper left portion of the screen as the place for the most essential elements, perhaps a visual, and the lower right portion for the second most important element, perhaps supporting text. The placement is dictated by what Rubens calls the "Golden Rectangle" (78). Using the upper left and lower right portions of the screen in this manner actually agrees with the principles of asymmetrical balance. The upper left portion of the screen can also be reserved for cues like flashing signals; as Rubens and Krull state, "Placement of text probably is the most effective technique for the static printed page. However, the dynamic screen may be more effectively exploited by selective variations from a consistent format" (32). Perhaps the best guideline for choosing how to indicate emphasis again rests on the principle of contrast as a source of meaning. Even the slightest contrast will give emphasis and capture attention.

When considering sequence, even the writer-designer bound by strict company format guidelines or a paucity of graphics might remember that type contributes to sequence. Legibility, ease in connecting letters and words that speeds eye travel across a page, can be enhanced by serif type (the type that rests on small "feet"). While some experts consider sans serif (without the "feet") as legible as serif type, a great many readers still prefer more traditional letter forms with serifs for reading ease (Benson 37). Most writer-designers prefer 9- to 12-point serif type because this size and style seem to contribute to sequence along a text line (see, e.g., Zoe Brown). Along with serifs, stress and thick-thin contrast in type contribute to legibility. As figure 14 shows, stress usually means that the bottoms of the letters lean slightly to the right; type styles with some con-

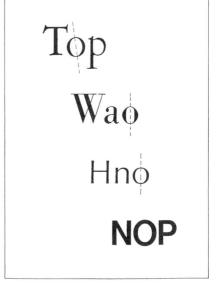

Fig. 14. **Popular Type Styles.** Garamond ("Top") has stress (seen by drawing a line from the thinnest top part of the small "o" to the thinnest bottom part of the "o"), serifs, and some contrast between the thick and thin strokes of letters for high legibility. Bodoni ("Wao") has no stress but distinct serifs and high thick-thin contrast. Optima ("Hno") has no stress or serifs and not enough contrast for high legibility in large text blocks. Helvetica ("NOP"), especially set in all caps, is acceptable only for display heads.

trast created by the thin and thick stroke of each letter have enough characteristic variation that the audience does not tire in moving from letter to letter. Capitals and lowercase, rather than all capitals, also contribute to legibility and therefore to sequence. At the least, the writer-designer can use type style to help the audience travel across the page. On the computer screen, however, stress and contrast are more difficult for the audience to see, and extremes in contrast and stress actually decrease legibility of letters on the screen (McVey 24).

Audience and purpose determine how sequence throughout the entire page or screen is controlled. The scanning audience needs additional help moving rapidly down the page or screen, from heading to heading, from unit to unit. Headings that spill over into the margin contribute to fast eye travel as well as to interesting proportions. The module approach to computer display design encourages the computer screen viewer to consider each screen a unit in sequence, with no backtracking necessary. Finally, the audience who must interpret, memorize, or enact each step in the documentation needs a visual sequence based on the content logic. Cautions or warnings in instructions must be displayed before the steps to which they apply, key words must be defined and emphasized, and lists must be indented and numbered or indicated by bullets. While considerations of audience and purpose are rhetorically important, they are also matters of design sequence.

In this discussion I have suggested *only* some of the applications of design principles to technical documents. When confronted with the task of setting a format, laying out a page, or creating a computer screen display, however, technical writer-designers can use design principles to make effective decisions. Knowing that unity, balance, proportion, and emphasis and sequence are essential in design and realizing their effects are the first steps; applying these principles to such choices as typography, size and placement of graphics, margin width, borders, alignments, and even bindings improves with practice and leads to an experienced eye for design. The technical writer now has the freedom and the challenge to develop that eye.[3]

NOTES

[1] One of the most useful texts on design principles, the one on which I have based many of my comments, is Lauer's *Design Basics*. For a concise definition of design principles as applied to advertising, see Roy Paul Nelson. While both Marra and Sadowski stress the importance of design principles in layout of technical material, they do not provide the detailed explanation of application that technical writers and designers now need.

[2] For a discussion of white space in display screens, see Bradford; Rubens; and Rubens and Krull. Although Rubens and I disagree about the overall importance of design principles in preparing texts, I suspect that he is unaware of his own use of these principles. For example, when he stresses that "specific areas of a page for specialized purposes can help the reader move more confidently through a text," he is employing the design principle of sequence, and when he recommends using certain areas for cuing on the computer screen, he applies the design principle of emphasis

(Rubens 79). For the other point of view, see Doty, who states, "Because the design of a page may influence whether or not your reader actually reads the document, page design is critical in shaping your document. . . . I suggest you [technical writer-designers] educate yourself thoroughly in the elements of good graphic design" (427).

³ For the reader who wishes to go beyond the works cited in this text, I would recommend highly the bibliographies in Pinelli, Cordle, and McCullough; Benson; Rubens; and Rubens and Krull. These bibliographies contain specialized works in both computer and page design.

Parallelism in Scientific and Technical Writing

ISABELLE KRAMER THOMPSON

TEXTBOOKS IN SCIENTIFIC AND technical writing offer the following statements about parallelism:

> Express items of the same importance in the same grammatical form. (Lannon 551)

> Make the elements in a series grammatically parallel. (Houp and Pearsall 461)

> Faulty parallelism results when joined elements are intended to serve equal grammatical functions but do not have equal grammatical form. (Brusaw et al. 483)

At first glance, this information appears to be helpful. A second look, however, reveals the lack of clear definitions. What do the expressions "items of the same importance," "equal grammatical form," and "grammatically parallel" mean? Does parallelism affect readability? Does parallelism occur frequently enough in scientific and technical writing for us to be concerned about its use? One approach to answering these and related questions is a close examination of corpora of well-written scientific and technical prose. In this article, I analyze two bodies of scientific and technical prose in order to speculate about the importance of parallelism as a stylistic device.

Description of the Study

The corpora examined include the eight feature articles from *Scientific American*'s June 1987 issue and most of the selections in Debra Journet and Julie Lepick Kling's *Readings for Technical Writers*. The articles from *Scientific American* contain approximately 34,000 words; the selections from *Readings for Technical Writers*, almost 26,000 words. Although the corpora are small, arguably both *Scientific American* and *Readings for Technical Writers* represent well-written scientific and technical prose. Possibly the most respected general science magazine available in the United States, *Scientific American* is known for sophisticated but comprehensive coverage of current scientific research. Journet and Kling made selections for *Readings for Technical Writers* with two goals in mind: "to provide a wide selection of the standard forms of technical writing in a variety of subjects" and "to present examples of effective technical writing" (1). Only technical instructions and specifications were omitted, because these types of writing, with their fragments and imperative sentences, show little stylistic variety.

By close analysis of these two corpora, totaling approximately 60,000 words, this study investigates how frequently parallelism occurs, what the most common types are, and how parallelism is used as a stylistic device. It also begins to consider some currently undescribed occurrences of what appear to be parallelism—for example, the sameness of form that sometimes occurs between sentences and among items in a list. Finally, this study examines faulty parallelism. Like other analyses of corpora, this study cannot claim to identify all types of parallelism found in scientific and technical prose; it can only show a wide range of the more common possibilities.

The Occurrence of Parallelism and the Most Common Types

To identify and classify the occurrences of parallelism within sentences, I used Mary Hiatt's scheme, devised for analyzing the parallelism in a 250,000-word corpus of varied types of writing (sampled from the 1,000,000-word Brown University corpus). Hiatt defines parallelism as "(1) the repetition of two or more words of the same form class in the same functional or syntactic situation, or (2) the repetition of two or more constructions of the same grammatical classification in the same functional or syntactic situtation" (17).

Based on this definition, she classifies each instance of parallelism according to the number of its constituents: doublets, triplets, and series. As the names suggest, a doublet consists of two parallel constituents; a triplet, of three; and a series, of four or more. Each type can be classified as simple (the constituents are usually single words with not more than one postmodifier) or complex (the constituents are constructions or single words with more than one postmodifier). After classifying an occurrence according to the number of constituents, Hiatt identifies any rhetorical schemes and repetition that accompany the parallelism.

Hiatt's classification includes all occurrences of parallelism, regardless of stylistic effect. Most occurrences are likely to be unnoticed by the reader, as shown in examples 1 and 2:

Example 1

We will also prepare financial pro formas based on projected income and expenses for the first and third full years of operation.

("RERC Proposal," *RTW* 149)

Example 2

The investigator shines a beam of light through a sample and measures how much of the light is absorbed.

("Molecular Mechanisms of Photosynthesis," *SA* 44)

In example 1, "projected income" and "expenses for the first and third full years of operation" are modified parallel nouns, which are classified as a complex doublet; "first" and "third" are parallel adjectives, which are classified as a simple doublet. In example 2, "shines a beam of light through a sample" and "measures how much of the light is absorbed" are parallel verb phrases, which are classified as a complex doublet.

The frequency of occurrence per 1,000 words of parallelism in the *Scientific American* corpus is shown in figure 1. Parallelism thus occurs approximately 23 times in 1,000 words, and the most frequent type by far is the doublet, accounting for more than 20 of the 23 occurrences. Rhetorical schemes, including antithesis, anaphora, homoeoteleuton, epistrophe, and repetition accompany 8.08 of the 23 occurrences.

Fig. 1. *Scientific American*

Type	Frequency
Doublets	20.88
Triplets	1.56
Series	.65
TOTAL	23.09

Fig. 2. *Readings for Technical Writers*

Type	Frequency
Doublets	29.29
Triplets	4.08
Series	2.24
TOTAL	35.61

The frequency of occurrence per 1,000 words of parallelism in the *Readings for Technical Writers* corpus is shown in figure 2. Occurring more than 35 times per 1,000 words, parallelism in the *Readings for Technical Writers* corpus is more frequent than in the *Scientific American* corpus. In *Readings for Technical Writers*, however, a lower percentage of the parallelism is accompanied by rhetorical schemes and repetition: 9.70 of the 35, slightly more than 1 in 4, compared with 8.08 of 23, slightly more than 1 in 3, in the *Scientific American* corpus. As in the *Scientific American* corpus, doublets account for most of the parallelism.

These findings fall close to Hiatt's. In her corpus, Hiatt considers what she calls imaginative and informative prose. Her samples of informative prose include news, sports, and society reporting, editorials, and reviews; writing about religion (including sermons); learned writing (academic articles); belles lettres (primarily essays); writing about skills and hobbies (how-to articles); writing about popular lore; and a miscellaneous category. The frequencies of occurrence per 1,000 words for parallelism in the nine categories of informative writing range from a low of 25.60 in newspaper reporting to a high of 44.48 in the miscellaneous category. Five of the nine categories report frequencies of occurrence from 33.23 to 36.95. The *Readings for Technical Writers* corpus falls into Hiatt's midrange; and the *Scientific American* corpus falls slightly below press reporting, Hiatt's category of informative prose with the least frequent parallelism.

Uses of Parallelism

Aided by cohesive devices that name the relation, parallelism directs readers to compare, contrast, or coordinate constituents of equal semantic weight in a sentence. In many cases, doublets are symmetrical with each other, bringing the balance and regularity that lend pleasure in art and architecture to writing (Graves 120; Leech 67). In other cases, parallel triplets and series increase in importance and build to a climax as the sequence ends (Sopher 45). The balance and regularity of parallelism help increase the clarity of writing; symmetry and climax add emphasis to call attention to certain constituents. To illustrate the uses of parallelism in scientific and technical writing, this section discusses a few of the sentences that occur in the corpora.

Parallelism can set up a comparison:

Example 3

Acceptance of the proposal also constitutes acceptance of these terms and conditions.   ("RERC Proposal," *RTW* 150)

The parallelism occurs before and after the verb, with the repeated "acceptance of." The two constituents are symmetrical with each other and equated by the verb "constitutes." Parallelism that sets up constituents to be compared or, as H. Sopher says, to be made equivalent with each other (46) is possible with "to be" and similar equalizing verbs.

Example 4 shows contrast:

Example 4

The child is able to infer that the vase is more like a plate or a rock, [both of] which fall, than it is like a bird or a balloon, [both of] which do not.   ("The Connection Machine," *SA* 113–14)

This sentence uses "than" to signal the contrast that has been set up by the parallelism. The two constituents are symmetrical with each other, and the balanced contrast (antithesis) is reinforced by repetition (anaphora). Regardless of the syntactic complexity, this sentence will probably be easy to read because of its predictable structure.

Example 5 shows a coordinated doublet:

Example 5

Yet neither of these phenomena involve[s] masses large enough and fast enough to generate appreciable waves.
("Gravitational Wave Observatories," *SA* 50)

"Large" is coordinated with "fast," and the parallel is emphasized by the repeated "enough" (epistrophe). Although the following version is slightly more concise, notice how much less emphatic the sentence becomes when "enough" is not repeated:

> Yet neither of these phenomena involves masses large and fast enough to generate appreciable waves.

In example 6, a series of items is coordinated:

Example 6

Screws, bolts, nuts, steel, packaging—virtually all kinds of products may be sold through representatives.
("Manufacturers' Sales Representatives," *RTW* 27)

The length (five items), the lack of a final conjunction (asyndeton), and the dash emphasize the series. Although some climaxing may occur as the series proceeds from small to large, the primary effect is to call attention to the number and range of products sold through representatives without pointing to a single product.

Example 7 shows other serial arrangements:

Example 7

Over the past 20 years we have identified neural structures and stations (large arrays of cells) that contribute to memory, traced their connections and tried to determine how they interact as memory is stored, retrieved or linked with other experience.   ("The Anatomy of Memory," *SA* 80)

The triplet of verbs in the main clause—"identified," "traced," and "tried" followed by "to determine"—is arranged in chronological order, beginning with the first task that researchers undertook and ending with the last. If the series is at all climactic, the emphasis on the last constituent occurs because it names the last task, not necessarily because that task is the most important or the most difficult. However, the second triplet, the series of verbs at the end of the sentence, seems to show climax. Emphasis builds as the functions of memory become more complex, from storage and retrieval to linkage. (For a full discussion of climax and emphasis in a series, see Weathers.)

Symmetry, or at least a sense of symmetry, can occur with constituents that are coordinated as well as with those that are compared or contrasted:

Example 8

A net result was that some 10,000 years ago the Northern Hemisphere received about 8 percent more solar radiation in the summer and about 8 percent less in the winter than it does now, making the summers

generally warmer and the winters generally colder than they are at present.
("Drought in Africa," *SA* 36)

"More" is juxtaposed with "less" and "summer" with "winter" in the first coordinated doublet (antithesis). "Summers" is juxtaposed with "winters" and "warmer" with "colder" in the second (antithesis). The repetition in both doublets reinforces the symmetry.

In the final pair of examples, regularity of style calls attention to irregularity of content:

Example 9

The sound of a familiar voice on the telephone summons a visual memory of the caller's face; the sight of a purple plum brings to mind its taste.
("Anatomy of Memory," *SA* 87)

Example 10

Seeing a familiar object, they cannot recall how it smells; after smelling it, they still cannot recall its taste.   ("Anatomy of Memory," *SA* 88)

In example 9, sound summons sight and then sight summons taste; in example 10, sight does not re-create smell and smell does not bring a memory of taste. The semicolons in both sentences emphasize the strange symmetry. The upset balance in these sentences comes close to Geoffrey Leech's notion of foregrounding, which occurs most frequently and purposefully in poetry.

As shown in the examples in this section, the parallelism that occurs in scientific and technical writing can be quite complex. Often reinforced by repetition and rhetorical schemes, it is varied and interesting. Regardless of its complexity, however, parallelism in scientific and technical writing is less noticeable than parallelism in political speeches and sales pitches or in poetry. These essentially oral media use parallelism to increase clarity by providing important signals to listeners. In such communication, parallelism reinforced with emphatic devices goes beyond its functional role of easing comprehension and assumes a more ornamental role of persuading or delighting. In poetry, parallelism goes to the extreme of calling attention to form as an aspect of meaning.

Parallelism in scientific and technical writing is primarily functional. Supposedly lacking ornamentation, practical writing is itself functional, concerned with informing rather than delighting and with persuading by providing information. As written rather than spoken communication, scientific and technical writing depends on parallelism to enhance the clarity of its often difficult content. The goal of parallelism in scientific and technical writing, therefore, is to increase readability—that is, to increase recall and reading speed. It points to information readers need to remember and identifies ideas they need to connect. The emphasis that results from infrequent types of parallelism or from reinforcing rhetorical schemes calls attention to the content the parallelism connects, not to form as ornament or meaning.

Other Considerations

Along with the types of parallelism for which adequate descriptions are available in expository prose and poetry, the corpora contain some seemingly parallel structures not often discussed. This intuitive parallelism occurs between dependent and independent clauses, between and among sentences, and among items in lists. Some of these occurrences are described in this section.

Parallelism between dependent and independent clauses

Since dependent and independent clauses are not structurally equivalent, they cannot be parallel, but a coordinated doublet or the repetition of syntactic structures and words to set up a comparison or contrast can make these unequal constructions seem parallel. The following example shows a dependent clause set up in comparison with an independent one:

Example 11

Gravitational waves differ from the more familiar static gravitational attraction in the same way that light and radio waves differ from static electricity and magnetism.   ("Gravitational Wave Observatories," *SA* 51)

Although the two clauses appear parallel because of their similar structure and because of the repeated "differ from," the content in the dependent clause is inconsequential to that in the independent clause. The clauses do not have equivalent semantic weight. What appears to be parallelism is limited to a similarity in surface structure.

The relation between the dependent and independent clauses in example 12 is more nearly equal:

Example 12

As the wave moves along the tube, it distorts the cross section into an ellipse; the major axis grows by a fraction of the original diameter, while the minor axis shrinks by the same fraction.
("Gravitational Wave Observatories," *SA* 50)

Even though the "while" makes the second clause dependent on the first, the two clauses after the semicolon seem symmetrical, with "major" juxtaposed against "minor" and "grows" juxtaposed against "shrinks." The similar phrases "by a fraction" and "by the same fraction" reinforce the symmetry. In addition, the content of the two clauses has equal semantic weight; neither clause depends on the other for meaning. Preceded by a comma, "while" seems to function in the same way as a coordinate conjunction. Notice how easily it can be replaced with "and" and a connecting word:

> As the wave moves along the tube, it distorts the cross section into an ellipse; the major axis grows by a fraction of the original diameter, and, simultaneously, the minor axis shrinks by the same fraction.

The parallelism here extends beyond the surface structure. The writer seems to intend the second constituent to be perceived as coordinate with the first rather than as subordinate to it.

Parallelism between sentences

At present, no description is available to classify occurrences of between-sentence parallelism. As Hiatt says, if the description of within-sentence parallelism is extended beyond the sentence, the criterion of repetitive constituents will declare every simple sentence parallel to every other simple sentence (22). Therefore, the criterion of repetitive constitutents is not exclusive or discriminating enough to identify occurrences of between-sentence parallelism. How much variability should be allowed between constituents? What should the required boundaries of their similarities be? A second problem is to characterize the relation that exists between parallel constituents in different sentences. Traditionally, structural relations have been confined to occurrences within sentence boundaries. Is the parallelism that occurs between sentences a structural relation? Are the two sentences linked by structural similarity, or is the relation between the sentences a result not of parallelism but of cohesion?

Even though an attempt to arrive at a description of between-sentence parallelism is beyond the scope of this study and impossible with such small corpora, I will examine a few examples. The occurrences of what appears to be between-sentence parallelism perform three functions in the corpora: setting up complementary or contrasting pairs, setting up a range without naming a list of items, and setting up a list of coordinated items.

Example 13 shows a pair of sentences that complement each other:

Example 13

The CPU is controlled by two kinds of programs. "Software" programs are entered by the machine operator, stored in the memory unit and can be changed as often as desired. "Firmware" programs are built into the systems, usually in Read Only Memory (ROM) devices that store instructions permanently.

("A Consumer's Guide to Personal Computers," *RTW* 24)

The beginnings of the second and third sentences are symmetrical: software and its accessibility are juxtaposed against firmware and its lack of accessibility.

Although inexact, the similarity of structure continues through the sentences, and the constituents carry equivalent semantic weight. Example 13 is comparable to example 14, which shows within-sentence parallelism, two parallel independent clauses separated by a semicolon:

Example 14

To generalize greatly, these designs fall into two broad classes: "coarse-grained" and "fine-grained." Coarse-grained machines link relatively few processors, each with a relatively large amount of computational power; fine-grained machines link a great many weak processors.

("The Connection Machine," *SA* 114)

The primary difference between examples 13 and 14 seems to be the length of the constituents. In example 13, the two constituents are perhaps too long for a single sentence; in example 14, they are shorter, so that a semicolon rather than a period can show their separation.

Example 15 shows two sentences that establish a range:

Example 15

These two classes of parallel computers form a spectrum. At one end is the conventional sequential computer, which has the minimum number of processors: one. At the other end of the spectrum are designs such as that of the Connection Machine, which include a very large number of small processors.   ("The Connection Machine," *SA* 114)

Along with equivalent semantic weight, these two sentences have similar structures and opposing or repeated words.

Example 16 shows a series of three coordinated sentences:

Example 16

For example, a line segment might be thought of as a "one-cube," or a cube with one dimension. Joining two one-cubes by their ends yields a two-cube, or a square. Joining two two-cubes by their corners yields a three-cube, which is what we ordinarily think of as a cube. Similarly, joining two three-cubes by their corners yields a four-cube.

("The Connection Machine," *SA* 111)

These three sentences are as closely related as the coordinated triplets in a within-sentence series. Compare example 16 with example 7.

In the next example two sentences set up coordinated items. Each sentence is followed by an explanation.

Example 17

In developing criteria to distinguish these two groups from the remainder of the Air Force population, we are guided by two considerations. First, we propose to use criteria that are concrete and minimally dependent on individual variations in attitudes and values. [two-sentence explanation] Second, we propose to use policy-relevant criteria. [three-sentence explanation]   ("Alcohol Problems," *RTW* 38)

Again the sentences share equivalent semantic weights, they have similar structures, and they use several of the same words. Conjunctive adverbs signal the connection.

Most attempts to use parallelism to connect sentences of equivalent semantic weight rely on similar structural patterns and the repetition of words. The similarity is most exact in the beginning of the sentence, up to and including the verb. The connection between constituents is often named with a conjunction. In addition, the connected sentences are usually introduced with a general statement to set up the comparison or the series. Even with these similarities, however, it is difficult to decide if parallelism between sentences is regular enough to be described or if its recognition will remain intuitive.

Parallelism in lists

One characteristic that distinguishes scientific and technical writing from expository writing is the use of numbered and indented lists. Forty-five such lists occur in the selections from *Readings for Technical Writers*. These lists seem to fit into two categories: those with parallel triplets or series as items and those with sentences as items.

Example 18 shows a sentence ending in a numbered list with parallel constituents:

Example 18

In 1978, the FTC proposed that Congress approve certain regulations that would require funeral homes to: 1) itemize prices of a service's various components; 2) give quotations of price via telephone; 3) require permission of the closest relative prior to embalming the deceased; and 4) display inexpensive caskets as well as costly ones.

("Financial Analysis of Service Corporation International," *RTW* 208)

The pattern in this example is comparable to that in example 7. It is one long sentence ending in a series of verbal phrases. Some selections in the corpora present lists with numbered but not indented constituents; others indent the constituents and use bold hyphens to separate them.

Not all the lists have exactly parallel constituents. Although the constituents of the list in example 19 are noun phrases, the phrases are not in precisely the same form:

Example 19

The malignant cells may spread to other parts of the body by 1) direct extension into adjacent tissue, 2) permeation along lymphatic vessels, 3) traveling in the lymph stream to the lymph nodes, 4) entering the blood circulation, and 5) invasion of the body cavity by diffusion.

("Cancer," *RTW* 17)

Making the parallelism exact is simple: change "direct extension" to "extending directly," "permeation" to "permeating," and "invasion of" to "invading." However, the changes may not be necessary, because the numbers mark the boundaries of the constituents as clearly as a comma and a repeated form would.

When the items in the list consist of sentences rather than clauses or single words, there is often no attempt at parallelism. Example 20 shows three items in a list with sentence constituents:

Example 20

The term *terrorism* as used in this Note assumes the following restrictions:
1. Terrorism refers to *contemporary* activity. Historical parallels. . . .
2. Terrorism is distinguished from terror, which is the rule by force and fear "from the top," i.e., by a dictatorial regime.
. . . .
4. Mere threats of violence are not terrorism, *unless* they emanate from a group that has already engaged in terrorist acts.

("Terrorists and Terrorism," *RTW* 21)

It appears that the typical admonishments of textbooks to list items in parallel form may not be closely followed by experienced writers. The primary goal seems to be making the constituents easy to distinguish from one another. As long as that goal is accomplished, by either numbering or indenting the items, variability among the constituents is allowed.

Faulty or Inappropriate Parallelism

In the corpora analyzed in this study, parallelism has been used effectively to represent constituents of equivalent semantic weight in similar syntactic forms. The resulting regularity of style has served as a "linguistic icon" (Sopher 48) to signal regularity of meaning. However, regularity of style can be unintentionally broken or used in misleading ways. Broken or misleading regularity is often called faulty parallelism. In addition, even if excessive parallelism is regular and not faulty, it can lead to an overly complex style. The attempt to put too many "linguistic icons" into one sentence, to show too many relations at one time, is an inappropriate use of parallelism. This section considers the problems that can occur from faulty and inappropriate parallelism.

Faulty parallelism

Faulty parallelism occurs when a writer fails to present juxtaposed, compared, or coordinated constituents in similar grammatical form or when a writer attempts to juxtapose, compare, or coordinate items that are not obviously equivalent semantically. To investigate the occurrence of faulty parallelism, I extended my corpora to include the work of students as well as experienced writers and examined a range of assignments totaling approximately 24,000 words written by sixty technical writing students during the 1986–87 academic year. These students are junior and senior undergraduates majoring in engineering, forestry, and agriculture. Although technical writing students generally read well and their language skills have improved considerably since they were freshmen, their writing is not as complex or carefully edited as the work of experienced writers. Looking through the *Scientific American* and *Readings for Technical Writers* corpora and examining the student writing, I found two kinds of faulty parallelism: syntactic and semantic.

Faulty parallelism at the syntactic level probably causes few problems for the reader. At worst, the reader may feel mildly irritated at the writer's carelessness. In addition, this type of faulty parallelism seems to occur seldom, appearing only about ten times in the 24,000-word student sample. In the following sentence coordinated constituents are not written in equivalent syntactic forms:

Example 21

The paper reviews the basics of induction motors, their uses, and how to gain better energy efficiency from them.

Written by the same student, example 22 shows a correlative without syntactically equivalent doublets:

Example 22

In fact, the induction motor can act either as a motor or a generator. . . .

Discussed in detail by most textbooks, faulty parallelism at the syntactic level can be easily corrected by an alert copy editor. As writers become more experienced and develop routines for accommodating the conventions of standard written American English, they avoid faulty syntactic parallelism without consciously investing intellectual energy, in much the same way that they avoid comma splices and subject-verb disagreements.

Faulty parallelism at the semantic level causes more difficulties for readers. It occurs when writers try to express in parallel forms ideas that are not clearly equivalent. In the students' writing and in the two corpora, faulty parallelism at the semantic level seems to result most frequently from carelessness about expression, often requiring readers to look back at the sentence to decide what the writer means. By investing extra time and energy, readers can determine the writer's intentions in examples 23 and 24:

Example 23

Some materials used as dielectrics are a vacuum, air, paper, glass, ceramics, plastic, and many liquids.

Example 24

We are all quite familiar with such commonplace materials as metals, ceramics, and wood. They have been used by man as standard materials for his weapons, tools, and construction purposes.

In both examples, the constituents in the series are parallel in form but not in kind. The first example mixes constituents of varying levels of generality—for example, "air," "ceramics," and "liquids" instead of "gases," "liquids," and "solids." The second mixes constituents of differing levels of specificity and with differing connotations—"weapons, tools, and construction purposes"—instead of "weapons, tools, and houses."

The faulty parallelism in example 25 will probably cause more difficulty for readers:

Example 25

Some of the problems of the TCAS system, such as lack of Mode C equipage, altimetry errors, intruder maneuvering, and the domino effect, have been discussed.

The constituents in the series are not all equivalent, nor are they all clearly problems. The writer has not been exact. The following revision increases the clarity:

Some of the problems of the TCAS system, such as the lack of Mode C equipage and the possibility of altimetry errors, of intruder maneuvering, and of bringing about the domino effect, have been discussed.

Taken from an article in the *Scientific American* corpus, example 26 presents an equally difficult task for readers:

Example 26

We taught monkeys to perform delayed nonmatching-to-sample with objects chosen from a pool of 40, each of them distinctive both visually and tactilely. The animals did the tasks both by sight and in the dark, where they had to rely on their sense of touch to distinguish the sample object from the new one.   ("The Anatomy of Memory," *SA* 87)

The second sentence pairs two unequal constituents—"by sight" and "in the dark"—and a correlative reinforces the false pairing. The sentence can be revised to coordinate constituents that are semantically equivalent:

The animals did the task by relying on sight and later, in the dark, by relying on touch.

This version pairs sight and touch, the parallel that readers predict based on common sense and on the clues given in the previous sentence.

Faulty parallelism at the semantic level can be more serious than the examples show. Besides resulting from careless expression, it may also come from muddled thinking, where writers pair and coordinate inappropriate constituents because they do not know what they want to say. In these cases, even the most careful readers cannot determine the writers' intentions. When it exists at all, textbook advice about faulty parallelism at the semantic level is usually not specific or helpful, so the remedy seems to lie in practice and experience. Yet, unlike parallelism at the syntactic level, parallelism at the semantic level never becomes routine.

Inappropriate parallelism

When parallelism is overused, sentence structure may become extremely complex and compressed. Too many ideas are expressed too briefly. Too much embedding confuses rather than clarifies relations. Surface structure becomes too compact for easy access to underlying meaning. As with faulty parallelism, readers can probably understand the writer's intentions, but they have to invest extra time and energy.

Although the parallelism in example 27 is not necessarily inappropriate, readers who are expected to decipher paragraphs of this complex prose may become overwhelmed:

Example 27

A meteorological drought is sometimes difficult to identify with any degree of reliability, in part because of the nature of the phenomenon and in part because meteorological and climatological information in many African countries has been available for only a few years or is of poor quality.   ("Drought in Africa," *SA* 34)

This 47-word sentence has three occurrences of parallelism, two of which are marked with rhetorical schemes. The information it expresses may be easier to understand if the sentence is divided into three sentences and the compound verb at the end is changed into two main clauses:

A meteorological drought is sometimes difficult to identify with any degree of reliability. One reason for the difficulty is the nature of the phenomenon. Another reason is that meteorological and climatological information in many African countries has been available for only a few years, or it may be of poor quality.

In this case, the simpler revision may be better because the sentence that comes before in the corpus consists of 29 words and the one that comes after consists of 45 words. It is difficult to decide whether the potential for elaborate parallelism encourages writers to compose long, complex sentences or whether the reverse is true—that writers prone to long, complex sentences are also overly fond of elaborate parallelism.

Suggestions for Future Research

At an average of more than 23 occurrences per 1,000 words in the *Scientific American* corpus and an average of more than 35 occurrences per 1,000 words in the *Readings for Technical Writers* corpus, parallelism appears frequently as a stylistic device in the sample of scientific and technical prose examined in this study. As Hiatt predicts, the most common form of parallelism is the doublet, with triplets and series composing only a small percentage of the occurrences examined.

Since it is a common stylistic device, parallelism probably usually goes unnoticed, a vessel shaped by the content it carries. In scientific and technical writing, it is most likely to be noticed when it is a form that occurs rarely or when it is reinforced by a rhetorical scheme or repetition. A structural relation, parallelism sets up constituents for readers to compare, contrast, or coordinate. It tells readers that constituents are related even though it cannot name the relation. The relation is named by a cohesive device, usually a conjunction or repeated words.

Parallelism is likely to affect comprehension and readability. The balance and regularity it brings to writing help determine the facts readers remember and the connections they perceive. By emphasizing certain ideas, parallelism highlights important content. It tells readers to pay attention to juxtaposed or compared ideas. In addition, and probably more important, parallelism influences coherence. Operating within sentences, parallelism sets up connections between and among details; operating between sentences and in lists, if such structural relations are possible beyond the sentence, it sets up connections among larger elements and aids in the comprehension of overall meaning. In both cases, parallelism enhances bottom-up processing of information. Finally, by acting as recurrent formatting (De Beaugrande 171–72), parallelism increases prediction by helping readers decide quickly what to expect next. It is likely that, when used effectively, parallelism increases reading speed and recall of information.

At present, however, no research exists to confirm these intuitions. At least three types of empirical studies are necessary to determine the effects of parallelism on readability:

1. An attitude survey to evaluate the importance experienced readers attach to parallelism and to elicit their subjective judgments about problems caused by faulty parallelism.
2. Experimental studies of sentence comprehension, as measured by reading speed and recall, similar to the investigations of active and passive constructions done in the 1960s (e.g., see Fodor and Garrett; Gough; Savin and Perchonock).
3. Experimental studies of discourse comprehension as it may be affected by parallelism within target sentences.

In addition, studies need to be made of larger corpora of scientific and technical writing. In scientific writing, the work of Charles Darwin, Albert Einstein, Lewis Thomas, John McPhee, and Stephen J. Gould comes to mind. In technical writing, studies could focus on the effective use of parallelism in different genres (technical definitions, descriptions, processes, and reports) and on specific constructions characteristic of technical discourse (not only lists but also textual headings in reports, descriptive titles of graphics, and entries in summaries). Studies of written products offer valuable information about what we should be teaching. Besides confirming or disputing the advice in our textbooks, they help us to construct a theory of good scientific and technical writing. Once developed, this theory will provide unified information about the characteristics of well-written prose.

Beyond Impressionism: Evaluating Causal Connections

Michael P. Jordan

Followers of Aristotle are reported to have denounced Galileo's discovery of the moons of Jupiter, on the grounds that they were not mentioned in Aristotle's treatise on the subject. These followers refused to look for themselves, knowing that there was nothing new to learn and that they would be wasting their time. Frustrated and perplexed, Galileo insisted that the moons were there—he had *seen* them! Surely, he argued, repeatable observations must take precedence over theoretically or intuitively derived conclusions.

Complementing Traditional Impressionistic Values

The long-established writing process has recently received new attention as a valuable aid for understanding and teaching technical writing. We must recognize, however, that the progressive development and improvement of a text is product-oriented: the writer continually develops the work until it becomes an acceptable finished product. In addition, pragmatically, it is the product—not the method of producing it—that is judged by readers, and general product criteria have evolved as the basis for this judgment.

Throughout the stages of the writing process, the writer considers many aspects of the writing in terms of such general criteria as:

Is the aim clear?
Is the aim meaningful and suitable for these readers in this forum?
Has the aim been achieved?
Is the organization sound and well-signaled?
Has the writing adequate cohesion?
Is the message clear to the intended readers?
Is the writing appropriately concise?
Are the tone and style appropriate for these readers in this forum?
Have conventions (e.g., spelling and punctuation) been followed?
Have requirements been met (e.g., for style and length)?

Such questions are invaluable aids in the writing process and in the evaluation of the end product, but they are limited by their generality and their resulting restriction to impressionistic values. Impressionistic evaluations are ideal for judging dogs, literature, zucchini, and beauty-pageant contestants, but since expository writing is a problem-solving process that results in a product to meet definable needs, standards, and practices, it would seem reasonable to seek to go beyond impressionistic judgment by determining well-defined systems of practice within established categories.

As an example, how do we judge when cohesion is adequate? On what basis beyond impressionism do we decide that a text lacks cohesion? How do skilled writers achieve cohesion? How do we help student writers achieve a desirable level of cohesion? What *is* cohesion? These are the types of questions we must be able to answer in order to progress from the general concepts of "continuity" and "flow" to more concrete and useful principles of language use. The purpose of this essay is to show how we might proceed beyond impressionistic values through empirical research based on multiexample corpora.

Topic and Approach

To demonstrate the empirical approach, I have chosen to provide a brief analysis of intersentential logical connection, because the common misunderstandings of this topic can lead to inappropriate criticisms of the final product. As even logical connection is an extremely large subject, I concentrate on relations of the well-known topos cause-effect, a small but significant aspect of cohesion in technical English, and sketch the techniques only in barest detail.

At the same time, this discussion provides a worked example of language research using a multiexample corpus for analysis and description. This approach is described in detail by my analysis in Couture's anthology on research about writing ("Close Cohesion"). The premise of this research method is that we can all learn from a detailed analysis of the products of skilled writers and editors—as long as we approach the task uninhibited by prescriptive rules of usage.

For this work, the main barrier to understanding is the notion that "vague" reference is bad and specifically that the substitute word *this* must be followed by a noun to clarify the referent. If you really wish to learn from language, you must be prepared to look—and to look with an open mind.

Implicit Connection and Background

On many occasions, there is no need for a specific signal to indicate the relation of cause-effect; the simple juxtaposition of material semantically related in this way is sufficient. We can see this in example 1:

Example 1

In the early evening of October 21, 1974, the 93,000 tons dwt tanker *Universe Leader* began taking on a cargo of crude oil at Gulf's Whiddy Island terminal. By the next morning, 2500 tons of it was in Bantry Bay instead of in the ship's tanks.

Someone had apparently overlooked an open valve for half an hour.

(*Pollution Monitor* Feb.-Mar. 1975: 13)

The cause of the oil spill is given in the final sentence, and readers can easily understand the relation even though there is no specific signal to indicate it. The writer could have used a verbal signal such as "This accident was caused by someone apparently . . . ," but the connection is clear without it. From the sense of the statements we know that only the second sentence enters into the cause-effect relation, and any attempted "clarification" of this would have been unnecessary and verbose.

Before we proceed, we need to understand something about cohesion in texts containing logical relations. Statements in the texts (in clauses, sentences, or paragraphs) are connected to other statements in binary pairs of "clause relations" as defined by Eugene Winter's work "Replacement as a Function of Repetition." These relations have also been analyzed by John Beekman and John Callow and by Robert Longacre. More recently, articles by William Mann and Sandra Thompson have extended the relations and our understanding of them within a system of rhetorical-structure theory. Although the number of relations in language is quite small, their combinations in micro- and macro-structures of text are probably infinite, and it is these that create the complexity of textual cohesion. Example 2 briefly illustrates that complexity (here and in other examples, the terms under discussion are italicized):

Example 2

Solid austenitic stainless steel, with its excellent corrosion resistance and high temperature strength, enables relatively thin wall designs to be used *and hence* becomes a close rival to the Cr-1/2 Mo steel.

(*Metal Construction* Dec. 1977: 557)

In this complicated sentence the subject is separated from the two main verbs, *enables* and *becomes*, by the phrase beginning with the logical subordinator *with*, and the compound verbs are logically connected by *hence*. Thus there is a cause-effect relation between the two clauses and one between the *with* phrase and the first of these two (see fig. 1).

We will now examine the major signaling techniques that make up the complexities of the cause-effect relation.

Fig. 1. Analysis of Example 2

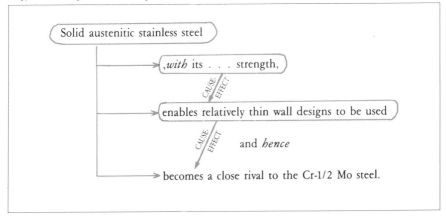

Logical Coordination and Disjuncts

Sidney Greenbaum's discussion of conjuncts, disjuncts, and adjuncts includes a semantic and syntactic analysis of logical connectors such as *so, for, therefore, hence,* and *accordingly.* Although *so* is sometimes used on its own as a sentence adjunct to connect sentences logically, most writers prefer to use it between clauses with *and*, as in example 3:

Example 3

The temperature of fuel stored in ships' bunkers or double bottom tanks approximates the sea temperature, which rarely exceeds 30°C, *and so* the SOLAS requirement is readily satisfied in respect of the 40°C flash point fuel.　　　　　(*Marine Engineering Digest* Apr. 1983: 19)

As a coordinator with *and, so* logically relates the two parts of the sentence that are coordinated, and so there is no confusion about which is the first part of the binary logical pair. *So* as an adjunct following a single clause and a semicolon also has this property; but when it is used as a sentence adjunct (like *therefore* but fixed in an initial position), the first part of the binary relation may be less obvious, as in example 4:

Example 4

The facts that you obtain from a survey with the refraction seismograph go hand-in-hand with the valuable information a soil or rock boring gives. Fewer borings, however, are required. *So* sub-surface surveys cost less money and take less time for both men and equipment.
　　　　　　(*Sealtest Catalog,* 1969: 41)

For this example, the first part of the cause-effect relation signaled by *so* is the difference *Fewer borings . . . are required,* with *however* mediating between the previous similarity and this difference. In other words, the "referent" for *so* (the part it is reentering into the new relation) is the second sentence, and the "referent" for *however* is the whole of the first sentence.

Recognition of the referential properties of all adjuncts is essential to an understanding of cohesion in language. The principle is often easier to recognize when we paraphrase the adjuncts in large expressions containing the substitute word *this.* As, in these paraphrases, *this* will have to refer to clausal referents, we can recognize that the adjuncts do so too. Although in some contexts *In addition to this* may lead to nominal or clausal interpretation, the *paraphrase* of *Moreover* in context cannot; that is, *Moreover* must have a clausal referent.

Adjuncts	*Paraphrases*
therefore	because of *this*
however	in contrast with *this*
consequently	as a consequence of *this*
moreover	in addition to *this*
alternatively	as an alternative to *this*

Once we accept that all adjuncts have clausal referents in the text, the fear or misunderstanding of clausal (or "vague") referents in text vanishes, leaving us free to examine more objectively the intricacies of textual cohesion. Look how naturally *because of this* works in an example of speech in an environment for which *therefore* is unsuitable:

Example 5

And a water molecule in particular can only join up, as it were, with another water molecule in a very special way. And it is *because of this* that you get these rather—er—special symmetrical patterns [of snow].
(*Dial a Scientist*, BBC Hatfield Polytechnic Transcript, Jan. 1975: 5)

"Vocabulary 3" Words

In "A Clause-Relational Approach to English Texts," Winter's analysis of words that directly "tell us" what the relation is between statements has led to his identification of "vocabulary 3" words, which supplement the conjuncts (vocabulary 1) and adjuncts (vocabulary 2). Vocabulary 3 words of special interest to us here are *cause, effect, result, consequence,* and *reason.* These tell us that a logical relation is present and often also indicate that it is one of cause-effect. Use of these words ranges from what John Hawkins has termed "untriggered" to fully triggered uses. The term *triggered* means that some mention of the referent is included with the new word (e.g., "The reasons for *this change*," "One effect of *this*"), whereas "untriggered" means there is no such mention (e.g., "The result," "The cause"). Example 6 illustrates an effect-cause relation with untriggered connection:

Example 6

Are you proud of how firmly you can hold a camera when you shoot at low shutter speeds? You still can't beat the blur problem, three Japanese scientists reported at a recent symposium of The Society of Photographic Scientists and Engineers. *The reason*, they say, is the vibration of the camera itself, even on a tripod.　　　　　　*(Popular Science* Oct. 1983: 4)

The referent for *The reason* is the main clause preceding the source and location clause in the second sentence. It is perfectly clear in this context, and any effort to "clarify" the referent would only make the writing less concise.

Strangely, few editors would question referent clarity in such examples, but the addition of *of this* often leads to requests to add a noun after *this*; but is "The reason for this inability" really any better than the original? The addition of *of this, for this*, and the like after the vocabulary 3 word is often a matter of stylistic softening rather than any serious attempt to clarify the referent. Note this and the clear clausal referent for *this* in example 7:

Example 7

In many instances, however, the application does not require even what we have called "low speed"; in fact, home appliances and many slow industrial processes are probably better served by the simplest, slowest integrating a.d. converter. There are *three reasons for this*: (a) much lower cost; (b) the much better noise rejection inherent in the integration process; and (c) the inherently higher stability of well designed integrating converters especially in industrial environments.
　　　　　　　　　　　　(Electronics and Power May 1977: 395)

There are several uses for attributive *this* (*this* followed by a noun), some of which are stressed comparison, naming, summarizing, classification, characterization, adding new information as an adjective preceding the noun, and—occasionally—referent clarity. Some roles of attributive *this* and the ways that writers achieve referent clarity with *this* or *that* are explained with many examples in my analysis of "The Principal Semantics of Nominal 'This' and 'That.' " The cline of markedness from untriggered to fully triggered vocabulary 3 logical signaling is shown in "Some Clause-Relational Associated Nominals in Technical English."

Rank-Shifted Connections

Once we accept the usefulness of "vague" referents, we can understand how writers condense or "rank-shift" clausal information into a concise nominal group. Instead of "These directional lighting effects result in video signals, which. . . ," we can simply write "The resulting video signals," as in example 8:

Example 8

Photomultipliers arranged in triads—groups containing one cell each for red, green and blue light—pick up light reflected by the model. These triads can be concentrated together to give directional lighting effects. *The resulting video signals* are passed to the display where a further pair of lasers produce light which is modulated to match the video signals before being combined. (*Canadian Controls and Instrumentation* Dec. 1977: 20)

The vocabulary 3 word *result* is used adjectivally to indicate cause-effect, and the cause is the untriggered directional lighting effects immediately preceding. Use of rank-shifted logical nominals is not a matter of trading clarity for conciseness. As the referent is perfectly clear without further embellishment, it can, in this context and in many others, be a valuable technique highly consistent with a mature, concise writing style. To test this, try rewriting the following passage in a more mature style without using a vague referent:

The company is to test an in-situ combustion process. This process burns part of the oil in a reservoir. The purpose of this burning is to heat the remaining oil. This heating in turn reduces the viscosity of the remaining oil. Finally this reduced viscosity enables the combustion gases to drive previously unrecoverable oil to a producing well.

The elegance of the original, example 9, is partly created by the use of the rank-shifted nominal, which is ideally suited to such chains of cause-effect relations.

Fig. 2. The Structure of Example 9

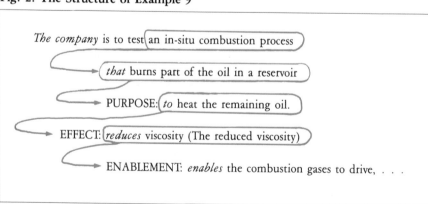

Example 9

The company is to test an in-situ combustion process that burns part of the oil in a reservoir to heat the remaining oil. *The reduced viscosity* enables the combustion gases to drive the previously unrecoverable oil to a producing well.   (*Chartered Mechanical Engineer* Dec. 1978: 25)

See figure 2 for an analysis of the information and structure of this example.

Verbal Connections

When using verbal connections, we need first to "reenter" a previous part of the text and then to attach it to a verb (often a vocabulary 3 word) within a new sentence. This is achieved in example 10:

Example 10

All lines of communication or interaction result in one common goal: to *improve productivity. Improved productivity* will inevitably result in cost control and cost management.
                              (*Engineering Digest* Sept. 1983: 35)

A previous part of the text (*improve productivity*) is reentered by full repetition, and the verbal form *will . . . result* indicates the cause-effect relation. It is not practical or even possible, however, to use repetition in this way very frequently, and writers use various techniques to reenter the reference clearly and economically. Example 11 illustrates a common method:

Example 11

When the dynamic stresses thus induced in the materials are of sufficient intensity local plastic deformation of the interface materials takes place so that oxides and other surface films are broken up and dispersed. *This results* in nascent metal contact, and a true metallurgical bond is formed in the solid state with no melting of the materials.
                              (*Fasteners* Jan. 1978: 33)

Note in example 11 the rank-shifted effect signaled by *thus*; the unsignaled cause-effect relation between the active and passive clauses of the second sentence; and, of particular interest to us here, the clarity of the clausal reference for *This*. There is no need to use "This breaking up and dispersal."

Where, however, there is a need to broaden the reference beyond the clause immediately preceding its reentry, we can use such techniques as *All this* or an epitomizing noun after *This*:

Example 12

The computer allows the program to be replayed on a computer terminal, *allowing* the programmer to see how the tool would move were it actually cutting a given part. *This approach* results in accurate NC tapes, minimal scrap and reduced prove-out time.

(*Engineering Dimensions* Jan./Feb. 1984: 26)

This on its own would have referred more readily to only the second clause of the previous sentence (a cause-effect relation signaled by *allowing*), but *This approach* broadens the reference to include both parts.

The *ing* verb form illustrated by example 12 is used when the two parts of the relation are quite short and are not connected to other ideas within the sentence. The usual alternative is *This* plus the verb form as part of a new sentence (*This allows* insteading of *allowing*). Try reading the second paragraph of example 13 using the *ing* form (*avoiding*) instead of *This avoids*:

Example 13

Foster Miller are to design a system that burns diesel fuel mixed with air to generate steam by direct contact with water. Both the steam and the combustion gases could be injected into the reservoir, *avoiding* problems of atmospheric pollution due to the combustion.

The Rocketdyne Division of Rockwell International are working on a similar idea in which the steam is generated in "downhole" heat exchangers, rather than by direct contact, and the exhaust gases are vented to atmosphere. *This avoids* the possibility of "plugging" the reservoir with particles generated during combustion of the fuel.

(*Chartered Mechanical Engineer* Dec. 1978: 25)

Variety is often a factor in using *ing* forms, and the technique is a valuable addition to novice writers' methods of logical connection. It is the larger pair of relations that is connected by *This allows* and the smaller pair by *allowing* in example 14.

Example 14

Additional test facilities are available from some suppliers offering frequency translation on the line output. *This allows* total modem testing when connected to a 2-wire line. Lower frequencies are translated up to the higher receiving frequencies *allowing* an effective analogue line loop to be achieved.   (*Telecommunications* Jul. 1977: 445)

Example 14 illustrates the two functions of *ing* clauses: logical connection between clauses (*allowing*) and descriptive connection defining or supplementing a noun (*offering*).

Although most logically connected *ing*-signaled relations are clear without further indications of the intended connection, *thus* and *thereby* are frequently used to reinforce the meaning. They are particularly useful when the verb in question is not a vocabulary 3 word, as shown in example 15:

Example 15

The TLP can also be untethered and moved to a different site, *thus saving millions of dollars in construction costs.* (*Compressed Air* May 1983: 15)

A further related technique, used for interconnected chains of cause-effect relations, is the clausal *which* (*which* with a clausal reference). Usually an *ing* form is preferred to the clausal *which* except in advertising where special emphasis is required, but use of an *ing* form is not always possible:

Example 16

Should the temperature of the system rise for some reason, the resistance of the heater increases, causing the heater wattage to drop, *which* results in a limit on the heater's temperature.
                                        (*Materials Engineering* Mar. 1983: 13)

The clausal *which* is ideal for such contexts, as another *ing* (*resulting*) could cause confusion and would represent questionable style.

Complex Combinations

We have already seen how the techniques described here are used not just in isolation but as a part of complex combinations of relations. Knowledge of the individual techniques should be seen as only a start toward understanding the cause-effect relation in technical English. Example 17 illustrates several techniques:

Example 17

Thus the drivers of the Bugattis, who knew perfectly well the lines on which they drove, only used their brakes when essential; for normal stops or decelerations they eased back their engines so that braking was only made from 60 kph (40 mph). *The time lost due to this* was made up by the acceleration at full power. *The reasons for these practical problems* are obvious enough.   (*Chartered Mechanical Engineer* Jan. 1983: 36)

Here we have a rank-shifted signal with the referent reentered by *this* and the cause-effect meaning given by *due to*. An alternative would be "Time was lost due to this easing back, but this time was made up by . . ." but a skilled writer would prefer the original. The last sentence in example 17 reenters the broad referent by *these practical problems*, indicates the cause-effect relation with the vocabulary 3 word *reasons*, characterizes these reasons as being *obvious*, and predicts details of the reasons to come. Such pivotal sentences are valuable cohesive links in language.

Here is a further complex example:

Example 18

It seems unlikely, *however*, that computer technology will transform the enduring characteristics of women's work [sic]. *On the contrary*, computer terminals in the home would probably *lower* secretarial wages. *The reason* is that the machines themselves make it possible to do large amounts of work quickly and *therefore* reduce the number of workers needed to do a given amount of work. *The resulting increase in competition for jobs* would drive prices down.   (*Scientific American* Sept. 1982: 176)

This paragraph is in a "contrast" relation with its predecessor (signaled by *however*), and the first sentence is the reference for *On the contrary*, which indicates "opposition" between the first two sentences. The first logical connection is in the second sentence, signaled by *lower*, and the second sentence is the referent for the untriggered *The reason*, which indicates the cause-effect relation between the second and third sentences; the two parts of the third sentence are logically connected by *therefore*. The part of the third sentence introduced by *therefore* is the reference for the rank-shifted subject of the fourth sentence, with *resulting* signaling the cause-effect relation.

Figure 3 offers an analysis of this example, including both logical and major lexical (word) connections. Readers should be able to draw such structural diagrams themselves from the examples and explanations offered here. If a text is open to more than one interpretation (can be drawn differently) or cannot be interpreted in clause-relational terms, we have grounds for criticism and a sound basis for improving the product.

We have seen here how a detailed analysis of cause-effect relations in practice can take us beyond an impressionistic understanding of this topos to a quite specific understanding of how these relations are communicated in English. It should be clear that this largely empirical approach can be applied to wider elements of "logic" in writing and that it can provide new insights into the broader concept of cohesion in English. However, a simple mechanistic knowledge of the signaling systems and the related semantics is still an inadequate basis for product evaluation. This essay has also attempted to provide some understanding of which method is better under given circumstances and what alternatives are available when one method has just been used. Effective product evaluation must go beyond psychology-based impressionism and rise above even a solid mechanistic language knowledge to a developed sensitivity of cohesion soundly based on systems of language use.

Just as we expect skilled cabinetmakers to use all the special tools at their disposal, so too we should expect technical writers to use all the signaling devices as appropriate. Ideally, writers should also be conscious of all the methods of logical connection and the best ways of using them. Those who

Fig. 3. Logical and Lexical Connections for Example 18

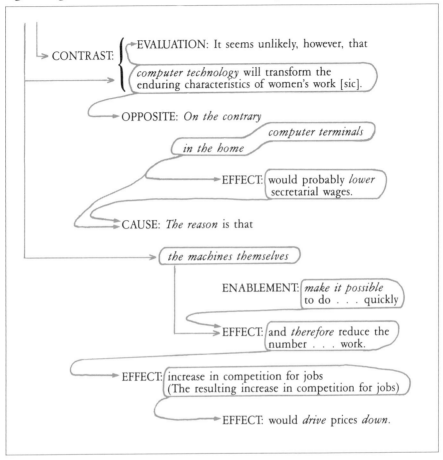

educate cabinetmakers should show their charges how and when to use all the specialist tools of the trade—even if they themselves find a couple of saws, a screwdriver, and a lump hammer quite sufficient for most tasks. Similarly, teachers of technical writing should show their students how and when to use all the signaling systems of English, even those they have a personal aversion to. Denial of any particular technique because some writers misuse it or because it violates an unexamined or misunderstood prescriptive rule detracts from true education.

In many ways, the approach outlined here is a return to the tradition of Aristotle, who studied the language of his day and offered theoretical principles and practical advice to contemporary communicators. Technical writing is considerably different from the public debate and disputation central to Aristotle's concern, and our educational mandate is far more diverse and challenging. Nevertheless, the task that faces us is quite similar to that of the ancient rhetoricians: to study the most important and best communications of the day, to describe the underlying principles and systems we find, and to use the resulting understanding as a sound basis for helping ourselves and our students to emulate and thus become skilled communicators. This task, though formidable, could eventually lead to a scholastically valid, language-based rhetoric of technical writing.

In earlier times, other followers of Aristotle felt that answers were to be found only in the minds or the published work of scholars. They blindly followed the edicts of their mentors, denying their own abilities to seek relevant evidence and to reason. And they rejected practical, repeatable evidence because it failed to coincide with established principle. Let us not make the same mistakes.

Galileo was forced to resign his professorship of mathematics at the University of Pisa after using empirical evidence to demonstrate that all falling bodies descend with equal speed. His book based on astronomical observations was banned, and he was called to Rome and made to recant his views.

As long as we are not blinded or inhibited by the half-truths of prescriptive rules and are not obligated through fear or ignorance to accept them uncritically, we can build our own telescopes and from our different perspectives look at the same truth together.

Further Uses of the Corpus

Once we have collected and used a corpus of examples for a specific purpose, we can use this basis to understand and explain almost any other feature of language use in communicative context. Although the corpus of eighteen examples included here is too small for any serious work on other features, it does show how a larger corpus would provide several informative examples of English use. Here are some of the language features exemplified in this corpus:

> How to combine actives, passives, and intransitives (examples 3, 4, 8, 11, 13, 14, 17)
> How to use possessives with inanimate objects (examples 1, 3, 16)
> How to use *which* with and without commas (examples 3, 8, 16)
> How to use *due to* (example 17)

We could also study punctuation, seeing how to use dashes (example 8), colons (examples 7 and 10), semicolons (examples 7 and 17), and quotations around words (example 13).

Preparing students to communicate effectively in their careers is the chief aim of technical writing courses. The timeless and timely issues of how best to accomplish this aim are subjects of the final section of this book.

In the first essay, two long-standing friends and colleagues, John H. Mitchell and Marion K. Smith, engage in the perennial debate over which is the better pedagogical model: prescriptive or heuristic teaching. While acknowledging that the prescriptive approach is no longer fashionable in pedagogical circles, Mitchell, a thirty-five-year veteran in teaching, argues that to dispense with this approach would be counterproductive. First, the forms of technical documents have become standardized throughout history and as a result are standardized throughout the Western world. This standardization allows communication to flow through time and across international boundaries. Second, within the United States, the style manuals of professional organizations are prescriptive, absolute, and often mutually exclusive. Both situations demand that we teach prescriptively. Not so, says Smith, who argues that the primary approach to teaching technical communication should be heuristic. Change is the most dominant characteristic of our age, and change demands that students be imaginative and creative. Echoing Carolyn Miller's argument in part 1, Smith maintains that the prescriptive approach implies utility, or low practicality; that it is reductive, making the writer an automaton of the technological world. On a higher level, the heuristic approach enables invention to reign over convention and imagination to triumph over standardization.

The second essay addresses the central problems for teachers of advanced technical writing courses: what to teach and what textbooks and materials to use. To help solve these problems, Thomas L. Warren divides his essay into four categories: manuals, proposals, visuals, and writing and editing. The texts he recommends for manuals and proposals are typically hybrid books that combine the process and the product approaches. Texts on visuals range from highly prescriptive to heuristic, as do those on writing and editing. He closes with helpful materials about printing and production, including desktop and mainframe publishing.

Counterbalancing James Souther's retrospective view in the first essay of the book, Elizabeth Tebeaux in the final and most futuristic of the essays looks toward preparing technical communicators for the twenty-first century, where rapid changes in communication technology will demand, not technical tactics, but insightful analytical skills and creative learning strategies. With the continued growth and accelerated transmission speed of information, future communicators must be able to develop and disseminate information to a wide range of users, including corporations, customers, and international audiences. The emphasis on analyzing audiences, designing lucid documents, and creating graphics will continue, but in new ways as information becomes increasingly complex, computer-mediated, and global. The implications for the future of teaching technical writing are no less than revolutionary.

The Prescriptive versus the Heuristic Approach in Teaching Technical Communication

JOHN H. MITCHELL AND

MARION K. SMITH

ALTHOUGH EXPERIENCED TEACHERS of technical writing differ on the relative merits of the two approaches, the prescriptive approach is recognized as the more efficient way of training students to perform in predictable situations while the heuristic approach is more suitable for educating students to perform in a changing world. In this two-part essay, the authors debate the pros and cons of each approach in the teaching of technical and professional writing.

Mitchell: The Prescriptive Approach

Courses in technical writing exist to train students to communicate in subject-matter areas and in the forms appropriate to those areas. Whether the goal is informational or persuasive, the forms of a document are set by professional societies, journal editors, military standards, and the handbooks and guidances of specific industries and agencies. Adherence to the shape, structure, and logic of these forms is essential to all professional-level writing and publishing. The forms are based on learning theory, information theory, and the concepts of readability analysis that are accepted internationally.

Although the phrase "prescriptive teaching" is no longer fashionable in the literature of pedagogy (say, of Hall; Stevenson; J. W. Allen; Winkler; Miller, "Invention"), I maintain that to delete entirely the prescriptive approach from technical communication courses would do our students an irreparable disservice. Therefore, I will present five rational and essential aspects of prescriptive teaching. I say "rational" because I do not intend to grow emotional and allude to great moments of humanism and literature. While I limit my comments to the teaching of technical and professional communication, some applications to teaching classical rhetoric or nonscientific writing will seem obvious.

This point of view is neither originally nor uniquely mine. It was introduced in a 1911 textbook by Samuel C. Earle. It reappeared in a 1917 textbook by Homer Andrew Watt. It was translated into educationese in the 1938 dissertation of Alvin M. Fountain. It was most eloquently articulated by Alfred North Whitehead in *The Aims of Education*:

> Finally there should grow the most austere of all mental qualities; I mean the sense for style. It is an aesthetic sense, based on admiration for the direct attainment of a foreseen end, simply and without waste. Style in art, style in literature, style in science, style in logic, style in practical execution have fundamentally the same aesthetic qualities, namely, attainment and restraint. . . .
>
> Style, in its finest sense, is the last acquirement of the educated mind; it is also the most useful. It pervades the whole being. The administrator with a sense for style hates waste; the engineer with a sense for style economizes his material; the artisan with a sense for style prefers good work. Style is the ultimate morality of mind. . . .
>
> Now style is the exclusive privilege of the expert. Whoever heard of the style of an amateur painter, of the style of an amateur poet? Style is always the product of specialist study; the peculiar contribution of specialism to culture. (19–20)

It is on this prerequisite of an educated mind and specialized study that the prescriptive teaching of technical communication is philosophically based. While I accept the argument of Carolyn Miller ("Humanistic") and other scholars that technical communication, especially scientific communication, can be persuasive, I maintain that the function of technical communication is to transfer discrete bits of information with minimum distortion, to decrease entropy. This transfer can be accomplished only when the carrier frequency selected by the transmitter is readable by the receiver. Education in predictable or standard patterns is essential if both transmitter and receiver are to deal with the same carrier frequency. In other words, the training pattern must ensure similarity. I suggest that, philosophically, specialist training must be prescriptive if one professional is to communicate with another in any subject-matter area, or "specialism," to use Whitehead's term.

History has borne out the philosophers, and that is my second point. The report of Frontinus to Caesar about the aqueducts of Rome ("Frontinus'") is strikingly similar to a report of Chicago's water commissioner to his city council. Specialists have learned to talk about their areas of specialty down through history. In fact, human beings were proceeding to heaven on a tower created with standardized communications. Construction was halted when nonstandard communications were introduced. I, for one, resent not being in heaven simply because technical communications became random and were not taught prescriptively. History also teaches that babel recurs when we fail to train prescriptively. Certainly the technological triumphs in the North Sea oil fields are directly traceable to the International Language for Servicing and Maintenance (ILSAM), just as the international acceptance of Caterpillar Corporation products was traceable to the prescriptiveness of Caterpillar Fundamental English (CFE). All reductive systems of language, including Ericsson English, are taught prescriptively.

Other international aspects require that technical communications be taught prescriptively. I have described the teaching of technical writing abroad, both in the United Kingdom and elsewhere. The data for these descriptions show that the major cultural entities of the West function with standardized communication formats. The late Horace Hockley, an officer in the Institute for Scientific and Technical Communication—London (ISTC), found the same phenomenon and made the reasons overt:

> . . . English is a compulsory subject in schools in many countries; it is the *exclusive technical language on the continent* [emphasis mine]; it is used by all airline pilots for communication with ground stations throughout the world; and it is the standard language of all standard organisations. This placed a particular responsibility on the ISTC to help other countries in the art of scientific and technical communication. (Hockley, qtd. by McIntosh 16)

Culturally, what this finding means is that the ISTC dominates non-American technical communication. Although the ISTC is not involved with teacher training, its Education and Training Committee generates the examinations for British certification as, variously, technical writer, author, or editor. These examinations are rigorous and professional. By ISTC definition, professionalism is a specialist form of culture that must be taught prescriptively if it is to be evaluated by a standard or by an external examination.

International acculturation extends beyond traffic control and certainly to military communications, which are particularly prescriptive among treaty organizations. For example, NATO's classified *Vista Report* and *Dictionary of Tactical Terms* control all NATO tanks and ships by using standardized communications. All non-Quebec airports control the world's air traffic with standardized English. Someone able to communicate within standards set by the American Standards Association needs no retraining to communicate within Australian, Canadian, or New Zealand standards. In effect, all specialized and professional communication worldwide is standardized, and, outside the United States, it is taught prescriptively.

Inside the United States, professional communication is also standardized. The domestic and international standards are often mutually exclusive or seemingly frivolous or arbitrary. (We are the only industrialized and English-speaking nation in the world that requires the spellings "liter" and "meter.") Almost all our administrative agencies attempt to standardize their communications. For example, there are *Guidances* for writers of environmental-impact statements. These *Guidances* vary with the subject matter, and it is impossible to teach the writing of environmental-impact statements without teaching prescriptively. One reason it cannot be done is that the Environmental Protection Agency has added *Guidances* for readers of environmental-impact statements. That is prescription squared.

Let me turn to my fourth point: the pedagogical reasons for teaching technical communications prescriptively. If we are honest, we will admit that most of us teach service courses for subject-matter areas. We do not teach content. Our students get their training in engineering, physics, whatever, from faculty members in other disciplinary empires. Those empires are jealously guarded; all are monitored by professional societies. It is a function of a professional society to issue journals. For example, the Institute for Electrical and Electronic Engineers issues thirty-three journals; the American Chemical Society issues fourteen. The point here is that each society issues a style manual. Those style manuals are Olympian; they govern everything that goes into the journals of a specific society. It goes without saying that they are arbitrary and prescriptive. They are also absolute. No societal journal will accept copy that is not consonant with that society's style guide.

Editors of specific journals have abdicated their editorial responsibility for form. They will reject out of hand an article of competent content and tell the author to align it with a specific style manual. It is the author's responsibility, not the editor's.

Equally important, many style guides are mutually exclusive in their prescriptiveness. For example, my university awards a PhD in physical chemistry. When graduates holding that degree write articles, they wind up schizoid. They must produce copy that is consonant with either the manual of the American Chemical Society or the manual of the American Institute of Physics. They cannot write an article acceptable to both, because citations, abbreviations, and section sequences are dissimilar.

Pedagogically, the situation is ideal. Given upperclass-undergraduate or graduate students (whoever heard of the style of an amateur), my textbook will be the style manual of the student's discipline. These students are specialists. They have identified with a discipline. They will welcome a text that is professional, prescriptive, and appropriate to their disciplines. Our task is easier precisely because of the professionalism of our students and the specialism of their areas. There is a rational basis for technical communication, and it is one of the reasons disciplinarians are winning the cultural war against permissive academics.

What we must continue to overcome is a pattern of misdirected heuristic training that began in the 1960s. For two decades, students in freshman composition were taught to write emotionally and to "discover themselves" in their own prose: "How do I know what I think until I see what I write?" Without rational standards and structures, they could achieve only an outpouring of self. "Any shrill scream was rewarded as a true high C," to use John Ciardi's comment on amateur poets.

Finally, and this is important, upperclass undergraduates recognize the fallacy of permissiveness. They see the illogic of "do your own thing" in prose. They realize that the world does not want the emotional outpourings of amateurs, and they insist that they be trained rationally and in their disciplines. As soon as they understand the essential humanity of intellectual honesty, they reject training without structure. And that leads me to my final point, the pragmatic basis for teaching technical communication prescriptively.

Technical communication is in the real world. It is goal-oriented and other-directed. In no way does it participate in the goal that preoccupied the CCCC in the sixties and seventies: the students' right to their own languages. Yet, even today, I must waste considerable time with students in an introductory technical communication class who have previously been heuristically taught. Students taught by freshman-English TAs schooled in literature are convinced the world wants to know their feelings, emotions, sensitivity, and opinions. If we do not prove to these students that in technical and scholarly communication only exact, reproducible data are important, they will fail as soon as they write a technical report or a professional article. The marketplace will reject our dilettantes.

We are honest, and our students are employable as technical communicators, if and only if we train them to communicate within the harness of a discipline. We are training professionals, and professional communications are governed by standards, guidances, style manuals, and specifications that fix the essential forms. Our students will survive as technical communicators only if they write in the forms, phrases, and symbology appropriate to the discipline or content area within which they work. If our students are to be employable, we must train them prescriptively.

Downstream are the results of current communications research into entropy and into aspects of quantum physics. It is too early to predict the application of this research to communication theory, but it is safe to assume that the application of the second law of thermodynamics and the theorems of Max Planck will be prescriptive and must be prescriptively taught. It is interesting to note that the work in entropy, called for by C. P. Snow back in 1959, is only now being done.

Smith: The Heuristic Approach

There is considerable merit in John Mitchell's argument that technical writing instructors should indoctrinate rather than educate, and I confess there are times when I conclude that it is the only doctrine that has a prayer of being cost-effective. But those times are the exception. I emerge from each grading session more convinced than ever that our long-term goals are best achieved by educating students to understand the why as well as the how of the communication situations they are likely to encounter. There is merit in enabling students to perpetuate the patterns of the past, but there is greater merit in enabling students to be curious and inventive. Awareness of prevailing patterns is important but not as important as immediate awareness of subject, audience, purpose, and situation.

There are essentially three reasons why our primary approach at the university level should be heuristic rather than prescriptive. The first is that there are strong limitations to what we can prescribe successfully. The second is that the prescriptive approach often creates or compounds the very problems it was intended to prevent. The third is that the heuristic approach—if understood and implemented properly—can enable our students to be more versatile and more useful over a longer span of time.

I doubt that the students we work with now are more intelligent than their predecessors, but I know that they have the means to be more independent. Automobiles, telephones, televisions, audio and video recordings, and other consequences of technological affluence have provided them with so many diverse models and so many opportunities to "do their own thing" that we must have very modest expectations of how successful we can be when we tell them what they should (or should not) do. The most dominant characteristic of our age is change, and our students have diminished respect for the past.

But even if we could be shrewd enough to persuade our students to do just as we tell them, it would be a mistake to train them that way. Conditions are changing too extensively and too rapidly for us to program students into fixed patterns of behavior. Ultimately, we must educate our students for the same reason that we put human pilots at the controls of commercial aircraft. Electronic technology is sophisticated enough now that there is little reason why a modern airliner cannot be fully automated to conduct all of its scheduled flights from the first step of takeoff to the last step of the landing. But our regard for human safety is such that we insist on having a human crew aboard because we have not yet mastered the art of programming machines to deal with the unexpected. No one wants to fly with an untrained pilot at the controls; if a major emergency occurs, we hope the aircrew will be able to add their own wisdom to what they were taught in flight school.

There is an axiom among soldiers that each war is fought at the beginning with the last war's weapons and tactics and that the ability to develop new ones is what will usually determine the outcome. The heat of battle seems needed to dissolve old patterns and allow new ones to flow into place. In World War II, for example, troops trained in the intricate arts of trench warfare found themselves ill-prepared to deal with the mobility of airborne assaults. Logistics as well as tactics suffered from a reluctance to abandon outdated practices. For many years enlisted men in the infantry were issued uniforms with six large buttons sewn inside the waistband of each set of "Class A" trousers. In compliance with military specifications, similar buttons were sewn into thousands of other trousers year after year—despite an army regulation that prohibited enlisted men from wearing the suspenders those buttons were meant to secure. The same reverence for established practice encouraged the French to build the Maginot Line and the American auto industry to build cars "longer, lower, and wider." Imaginative invaders found more effective ways of doing things and rolled on to victory.

I do not wish to downgrade the value of standardization. Auto mechanics of every level of expertise can appreciate the marvel of being able to go to independent dealers and buy exact replacement parts for automobiles built years ago and thousands of miles away. Frequent travelers are greatly convenienced by finding telephones and directories so similar to ones at home that they need to spend little or no time learning to use them efficiently. But standardization should be regarded as a means to an end and not as a goal in itself.

Let us grant that there is a broad range of established patterns in written and spoken communications that must be maintained to allow lateral communication among present communities and vertical communication from the past to the present and from the present to the future. There is even merit in maintaining inefficient systems simply because they are already in place (a point well-illustrated by the need to keep building railroad tracks with rails only 4 ft. 8½ in. apart even though a modern train drawn by a 500-ton locomotive could operate more safely and efficiently on rails eight to ten feet apart). The real or imagined cost of change can often be so great that sticking to an established but inferior practice would be the lesser of two evils.

But standardization for its own sake too easily becomes the prevailing policy and easily leads to stagnation. The layout of the computer keyboard this essay is being written on was deliberately designed to limit the speed of the user (to accommodate the limitations of primitive typewriters whose keys would jam under the hands of a speedy typist), but the standard is so universally honored that there seems little probability that we will see a conversion to the Dvorak layout, which offers far greater logic and efficiency.

Prescription is obviously necessary for the success of such artificially developed languages as Caterpillar Fundamental English (a variation of Basic English developed by the Caterpillar Corporation as a practical alternative to providing service manuals in all the native languages used in its worldwide construction projects). But Caterpillar English was not developed by someone who was content to follow prescribed patterns and practices. It came about because someone could visualize a better way of doing things, and it is likely to remain a better way of doing things only to the degree that its advocates have a clear understanding of its strengths and limitations. Perhaps some of nature's caterpillars can be useful here. I am told that tent caterpillars have a pronounced tendency to follow each other and will normally form chains while foraging for food, each one in the chain trusting the one ahead to determine the direction and rate of travel. All goes well as long as the leader keeps on the lookout for fresh feeding grounds. But if that lead caterpillar should ever happen to fall in behind the last caterpillar in its wanderings, the whole chain may be on the path to extinction.

One does not need to visit a biological laboratory to find tent caterpillars on the march. While doing some in-house training for a midsize industrial firm a few years ago, I asked to see samples of the major products of their working hours to get a clearer picture of what training would be most needed. When we came to their proposals, I found some huge documents that seemed all out of proportion to the worth of the goods and services the firm was marketing. Asked how they could afford to write such lengthy proposals for projects offering only marginal profits, the chief writer assured me that they did not write a completely new proposal for each project—what they did was to keep a few large boilerplate proposals on file that were reproduced with slight modifications to allow for descriptions of the particular hardware or service to be provided. Because the boilerplate models were so large, it had been years since anyone had taken a critical look at any part of them beyond the two or three pages that were routinely modified. The result was that the company was regularly sending out proposals designed many years earlier for a significantly different line of products and services. Buried in the musty pages of those ponderous proposals were commitments to send staffs of technicians to assist in integrating purchased equipment—equipment having a total value lower than the cost of providing the promised technicians. Apparently the only thing that had saved them from major embarrassment and possible financial loss was that the people receiving the proposals, accustomed to reading only the few updated pages, were unaware of substantial benefits they could have legally claimed.

A more dramatic example of standardization drifting into stagnation came from the combat experience of a well-known Utah journalist. While on temporary assignment to cover America's military involvement in Vietnam, Hack Miller managed to get aboard a helicopter conveying American troops into a combat zone. Just as they had cleared the trees and were descending into the designated landing zone, they spotted an enemy gunner on the ground—holding them in the sights of his automatic weapon at point-blank range. Just as Miller was deciding that his writing career had only a couple of seconds remaining, the Vietnamese inexplicably swung his weapon away and began firing at a point in front of the chopper. The ship's gunner immediately dispatched him with a burst of his own and they landed safely, but the crew was puzzled as to the dead man's motives. While they were examining his body, they found that he had been trained to deal with many standard situations. Laid out in full view was a card explaining the procedure for shooting at aircraft—and the card stressed the need to lead the target by several lengths so that bullets and aircraft would arrive at the same point at the same time—fine instructions for the propeller-driven aircraft used when the card was printed, but not much use for a hovering helicopter.

Both are extreme cases, to be sure, and may be more useful as metaphors than as examples of the problem at hand. Less striking instances are normally defended as organizational policy or as a proper response to an assumed obligation. For these, we may consider how intractably some individuals and organizations have continued to prohibit the use of first person by an author or how tenaciously they have clung to the practice of requiring elaborate footnotes long after the convenience and efficiency of parenthetical documentation have been amply demonstrated. I feel that helicopter hovering overhead when I see a college senior dutifully attempting to construct a thesis statement for a set of directions or for a description of a device—not because a thesis statement would be functional for such a purpose but because years earlier a teacher of the 500-word essay effectively indoctrinated this student. The persisting reluctance on the part of individuals and organizations to use graphic illustration, to integrate the graphics with the text, and to add color to the graphics can be attributed to a lack of skill or to a lack of technical facilities available to the writer, but we will do well to consider how much it has been encouraged by uncritical dependence on existing models.

It is true that we are expected to use the prescribed format when submitting an article to any journal published by the larger professional societies. But, even here, standardization proves to be a sword that cuts both ways. Almost all the major societies insist on a standardized format, but the standard exists within rather than among the organizations. Each one has so jealously and rigidly perpetuated its own policies that a genuine common market of information exchange is hindered as much as helped by the individual standards designed to expedite the flow of information. In earlier years such standardization posed little problem. Technical disciplines were few, rather easily defined, and relatively slow to change. But conditions are strikingly different now.

Increasingly, scientific and technological developments are rapidly erasing old boundaries and creating new ones as established disciplines merge, multiply, or attempt to coexist with new arrivals. Some disciplines (such as drafting) are either disappearing or are being so drastically redefined as to be virtually unintelligible from the perspective of those who created the governing standards. New disciplines (such as microelectronics, robotics, biological engineering, and ergonomics) continue to materialize at an impressive rate. Distinctions among disciplines continue to blur with the advent of new perspectives, methodologies, and applications.

Given such a dynamic state of affairs, what will govern the creation of standards used in our professional publications? Will the yet-to-be-founded journals of robotics, for example, be written in accordance with the style specifications honored by mechanical engineers rather than with those honored by electronic engineers? Or will that disputed territory be governed by the APA format instead because the American Psychological Association comes to be regarded as the primary authority on problems encountered when robots are equipped with artificial intelligence? Will biological engineers form a homogeneous group, or will genetic engineers march to the beat of a different drum than specialists in organ regeneration? Technical writers and editors of the future will need to be very imaginative indeed to correlate new discoveries with the knowledge and practices of the past.

Stylistic standardization is desirable (and particularly useful for routine communications that lend themselves well to the use of blank-form reports), but foresight and imagination are needed to break down the barriers so that information of joint interest can migrate from one discipline to another without hindrance. Earlier, Mitchell observed that standardization has made it impossible for a doctor of physical chemistry to write to both physicists and chemists in the same professional article and that we are now expected to have specialized training so that we can read environmental-impact statements. In making those observations, he reveals enough of the drawbacks of the prescriptive approach that I would submit my case to an impartial jury on those points alone.

Conventions rarely become established without at least appearing to serve some useful function, but they tend to outlive their usefulness. Our letters, for example, still commonly begin by stating that the intended reader (often a total stranger) is "dear" to us, and they end by claiming that we somehow (truly, respectfully, cordially, sincerely, etc.) belong to the person addressed. Why do we perpetuate such nonsense? It certainly falls far short of achieving the accuracy and precision we ask our students to aim at. We feel comfortable with it only because it is customary, and thus it offers the security of a prescribed pattern.

Our task as teachers is to educate students to have a healthy respect for prevailing practices, professional standards, established formats, and conventions in general without allowing that respect to harden into blind religious reverence. Prescriptions issued by a medical doctor will not be honored by a legitimate pharmacy after they become outdated, but prescriptions given to students by writing instructors have no built-in termination date. They tend to be uncritically honored and acted on long after circumstances have significantly altered their relations to relevant reality.

We must educate our students to be alert enough, imaginative enough, and flexible enough that they can see these things as means to ends rather than as ends in themselves and can abandon or modify means that no longer lead to the intended goal. A properly educated student will be able to size up a communication opportunity with an appropriate awareness of subject, audience, purpose, and situation. Having done so, that student should be able to make a defensible decision whether to follow a beaten path to the objective or to pioneer a new trail.

An intelligent application of the heuristic approach is not a retreat into the undisciplined "turn 'em on and turn 'em loose" philosophy that unfortunately flourished in the 1960s and early 1970s. It is a sober realization that students will need to function in a rapidly changing world where procedures, policies, and practices prescribed by veterans of earlier battles must be respected but not uncritically trusted to be relevant or reliable.

The issue is essentially a choice between tactics and strategy (a choice that can easily pit utility against humanity). At the tactical level, where we are driven by deadlines to win wars, build seaports, or construct powerplants, our immediate needs become our primary needs and we will do well to prescribe as energetically as we conscript civilians and train them to be interchangeable parts in a military machine. Under such conditions students are reduced to being extensions of technology itself, little more than mobile tools for getting a job done quickly and efficiently—but the job does get done in ways that allow immediate and obvious benefits to society.

The heuristic approach is more applicable at the strategic level, where invention outweighs convention and imagination is worth more than standardization. If we hope to build a better world, we will do well to regard the education of the student as a proper goal in its own right and to allow for the variables that make that student a unique human being with the capacity to make original contributions, independent judgments, and unbiased evaluations.

A Comparative Analysis of Teacher Resources for an Advanced Technical Writing Course

Thomas L. Warren

THIS ESSAY PRESENTS resources teachers can use to design an advanced technical writing course, preparing students to become communicators in business, industry, or government. In addition to learning fundamental communication skills in introductory and intermediate technical writing classes, students need to enter an organization prepared to work. The advanced course, then, is a professional, job-related course where students prepare for a career.

Text selection for advanced technical writing is different from text selection for earlier courses because there are no specific texts. Teachers must find materials that are not traditional texts—that is, they have no apparatus or teacher's manual. (For additional resources, see Moran and Journet; Warren; and Lindemann.) To help that process, I have divided the discussion into four areas: manuals, proposals, visuals, and writing and editing.

Manuals

We can divide potential texts for teaching students how to write manuals into four groups, evolving from the rhetorical form of instruction. I focus on product manuals, one of several specialized areas. (For books focusing on other areas, see n. 1.)

Writing product manuals

Donald Cunningham and Gerald Cohen's *Creating Technical Manuals* thoroughly explains how to produce a product manual. The "User-Friendly" of the subtitle reflects as much on the book and the student as it does on the product user. For example, product manuals consist of many parts that are unfamiliar to students (warnings and cautions; parts lists; specialized sections on installation, troubleshooting, maintenance; and the like). Students rarely understand the complexity of the contemporary manual and so need as much help as possible.

The authors guide students through putting together a manual, from developing a technique to get the writing done to explaining what the product is and does to repairing the product. The last chapter focuses on computer manuals.

Writing skills for computer scientists

Books for computer scientists are much more common than those that address the general problems of manual writing. Both William Skees's *Writing Handbook for Computer Professionals* and Charles Sides's *How to Write Papers and Reports about Computer Technology* describe the writing situation for computer specialists, stress the importance of good communication skills, offer suggestions on planning and writing various reports and memos, and close with comments on specialized aspects of writing.

Skees, a specialist in computer-related writing, gives students an in-depth analysis of what goes on when someone has to write reports, updates, and manuals about computers. Students follow the process and then get details about various kinds of writing: "Technical Writing," "Management Writing," and "Promotional Writing." Skees covers the writing demands of business, industry, and government that many technical writing textbooks ignore. While advanced students are presumably competent in writing and researching skills, they need information on what a manual is, how to put one together, and what happens on the job.

A writing teacher, Sides leads the student through preparing material from defining audience to inserting appropriate visuals. He also includes material on oral presentations and group conferences. That he teaches writing suggests the strength of the book.

Writing computer manuals

Susan Grimm approaches *How to Write Computer Manuals for Users* in a systematic way, based on her experience as a writer of computer manuals. She organizes the material around the before, during, and after elements in writing a manual. Her two sections on after the writing ("After You Write" and "After the Manual Is Finished") present often overlooked materials. Once the job is done, is it really done or is there more? Grimm points out that the manual-out-the-door is but the beginning because of the updates, revisions, and—an aspect few technical writing students are aware of—the customer-training sessions.

Structuring documents according to a definite plan is the focus of Sandra Pakin's *Documentation Development Methodology*. She emphasizes the importance of the plan (the DDM) more than the other authors (7 of her 16 chapters are on planning). The book's strength lies in the DDM's evolving from her work as a consultant (Sandra Pakin and Associates). Students can also learn a great deal about how to write documentation by studying her style and will learn even more by studying her use of layout and design to present the material.

In *Writing and Designing Operator Manuals*, Gretchen Schoff and Patricia

Robinson focus on books that the user reaches for when things go wrong with the system. Their book introduces another element that most advanced students are not aware of: product liability and the role documentation plays. Certainly, if something goes wrong with the product, investigators will look at the manuals as carefully as they look at the system. Schoff and Robinson also include chapters on preparing manuals for the international market and on managing and supervising manual production. Students need an introduction to both topics because many will work for multinational corporations and all should aspire to move into management positions.

Christine Browning's *Guide to Effective Software Technical Writing* traces the development of a computer-software manual from the organization of the manual through what constitutes front and back matter (i.e., prefatory materials, such as tables of contents and illustrations, and appendix materials). Browning assumes that the reader already has writing skills and wants to learn the specifics of software documentation.

G. Prentice Hastings and Kathryn J. King's *Creating Effective Documentation for Computer Programs* carries the advanced student through all phases of creating the documentation. What separates this text from others is the chapter on production scheduling and the frequent boxes containing key points. Both authors have experience in writing manuals.

Proposals

Proposals, like manuals, have some connection to the introductory and intermediate technical writing courses. Students in these courses routinely write some type of proposal for a project, so the teacher of the advanced course can assume the students have some, albeit limited, understanding of proposals. There is little resemblance, however, between the proposal taught in earlier courses and one for the job. Fortunately, there are several good books available.

Most government business, regardless of level, comes through outside providers. These vendors submit proposals to supply goods and services ranging in value from a few dollars to many millions. Two current books address the central issues in getting government contracts: Herman Holtz and Terry Schmidt's *The Winning Proposal* and Donald Helgeson's *Writing Technical Proposals That Win Contracts*. Both books make clear that successful proposals begin not on the day when the request for proposal (RFP) comes out but rather months before. Proposal teams need to be in place before the official start time so that the bid–no bid decision is not a hasty one. When the RFP comes out (or the announcement is made in the *Commerce Business Daily*), management can bring a team together quickly.

Holtz and Schmidt present an insider's view of the proposal process. Their approach is not the "I've been there before" approach that Helgeson uses, although both authors are experienced proposal writers. Rather, they discuss the pre-RFP activities, bid–no bid decisions, scheduling the process including the budget, and getting the document out the door. Like Helgeson, they devote little space to the mechanics of writing (1 out of 16 chapters in Holtz and Schmidt and 1 out of 7 chapters in Helgeson). Both books, therefore, concentrate on matters students need to know when working as proposal managers or as members of the proposal team.

Helgeson has spent many years as a proposal manager, so he brings considerable insight and numerous anecdotes to his book. His informal approach masks an urgent message, one taken from the Boy Scouts: "Be Prepared." He stresses the prebid activities of the manager, such as getting the marketing staff to develop a thorough profile of the current contract holder and any potential competition. Finding the soft spot in the competition is as much a part of audience analysis as understanding the thinking of those in the government who write the statement of work (SOW) and the government's contracting officer (a combination accountant and lawyer).

The proposal manager must have authority and access to key personnel and information. He or she must be able to plan carefully and budget parsimoniously because some proposals run into hundreds of thousands of dollars and countless hours of work. Helgeson also discusses the government's evaluation process, providing insider hints about how proposals are evaluated.

Technical writers also prepare proposals for business and industry, and while there are some differences, the approach is generally the same. (For books on nongovernment proposals, see n. 1.)

Visuals

Visuals (or graphics) offer a different challenge to the teacher of advanced technical writing. Whereas manuals and proposals are "genres," having distinct characteristics and standing alone, visuals are an integral and supportive part of almost every technical document. The teacher can build on what the student already knows by teaching in greater detail such matters as scaling, complexity, and different reproduction techniques. Then, the teacher can present different forms of visuals, such as the log-log chart, the semilog chart, and the numograph.

A. J. MacGregor's *Graphics Simplified* takes the student through a variety of visual forms. All the familiar forms are there (bar, column, and pie charts; line graphs; and others), but so are more intricate forms, such as surface and multiple surface curves, deviation and 100% column charts, paired and deviated bar charts, and many more. McGregor explains when a particular form of visual is appropriate and when not.

Robert Lefferts's *How to Prepare Charts and Graphs for Effective Reports* offers detailed discussion of the principles of visual design (Lefferts lists nine: clarity, simplicity, emphasis, summarization, reinforcement, interest, impact, credibility, and coherence). He then applies these principles to seven types of charts:

bar, column, pie, line, organization, flow, and time. For example, he discusses how shading affects bar and column charts and how the writer can use it to special effect in conveying the message.

Henning Nelms covers much more in *Thinking with a Pencil*. Nelms's thesis is that you can generate supplemental visuals that move beyond standard forms. His book really introduces students to more artistic forms—drawings, rendering, lettering, and so forth—giving the students information they need to produce visuals that artists can transform into finished products.

Last is a book by Edward Fry, the man who developed the reading formula and curve that carries his name. He aims *Graphical Comprehension* at a younger age group than the one our students belong to. But, because even advanced students are relatively unsophisticated visually, his book is still useful. It discusses a wide range of visuals from traditional diagrams and the various charts (bar, column, pie) to dimensional graphs, schematic drawings, and concept illustration. Teachers will find familiar ground in discussions of illustrated directions, process charts, and hierarchical charts.

All the books discussed so far address the issue of visuals as if the writer will either produce a final and finished copy of the visual personally or enlist the help of an artist. None of them discusses the problems and the solutions offered by computer visuals. Certainly programs that produce visuals (such as *PFS:Graph*) make drawing charts and graphs by hand unnecessary. But these programs introduce a series of new problems while solving the technical ones. For example, working on-screen distorts the writer's sense of perspective because of screen size and color. True, the writer can print out samples, but printers (unless laser printers) work slowly. Another problem is that some of the older visual programs do not allow easy access when the writer returns to the project. Users of *Lotus 1-2-3*, for example, have a difficult time reviewing graphs on-screen once they leave the spread sheet. The teacher, therefore, had best consult recent articles (listed in current bibliographies—see n. 1).

Writing and Editing

The final category, writing and editing, is such a nebulous topic (because of the many views, first of all, of what is good writing and, second, of how to edit efficiently) that resources are equally vague and hard to pin down. Consider that students, on the job, must edit not only their own work but also the work of others. How do technical writers explain to an accountant that the material in the cost-effectiveness report is obscured by the accountant's failure to understand what the reader needs? How, once that problem is solved, do technical writers prepare the manuscript to be typeset and then readied for printing? Both copy and production editing are areas our students need help in.

There are books specifically related to style (such as John Trimble and, of course, William Strunk and E. B. White); books that deal with writing in general and style (such as Thomas Kane); technical writing books that present style as the central point (such as James DeGeorge, Gary Olson, and Richard Ray); books that focus on editing your own work and, by analogy, the work of others (Claire Cook). There are also narrowly conceived books (such as those by Daniel Felker et al. from the Document Design Center) that focus on specific issues and subjects. The point is that the materials you use will depend on what your students already know. Do the textbooks your students used in earlier classes approach writing and editing with an intrinsic (some say innate) sense of style? Can teachers teach style to technical writers on a level other than mechanical?

Kenneth Houp and Thomas Pearsall's *Reporting Technical Information* discusses style as part of the audience-analysis process, but the considerations are really rather mechanical: a percentage of S-V-O sentences, so many words per sentence and per paragraph, level of diction, and method of organization. Any one of these topics could constitute the unit on writing in the advanced course. For example, in *Style and Proportion*, Josephine Miles shows how mechanical counting can reveal interesting things about the fiction, nonfiction, and poetry of many writers. In *Designs in Prose*, Walter Nash takes the counting process a little further by discussing the structure and design of prose, and Rodney Huddleston, in *The Sentence in Written English*, produces a syntactic study based on scientific texts (the study, by the way, is meant for linguists rather than for technical writing teachers).

If you prefer the mechanical approach, there are also computer programs that will do such analyses (such as Jim Button's *PC-Style*, Micro's *Readability Calculations according to Nine Formulas*, Reference Software's *Grammatik III*, and Bell Laboratories' *Writer's Workbench*). The technical writing teacher needs to be familiar with such stylistic-analysis programs because many organizations are currently using them or want to use them. Another part of style is company style. Most students are familiar with the style demonstrated and taught by the author of earlier texts. Perhaps they know the MLA system. When they graduate and work for a company, they will find a company style and, if they want to publish their work in a journal, a journal style. Various disciplines have either their own style manual—such as the *CBE Style Manual*—or one that they accept—such as the widely accepted *Publication Manual of the American Psychological Association*. (See Howell for a complete listing of such style manuals). Companies, however, frequently want things done their way and refuse to adopt outside style manuals. In many cases, style is a matter of tradition, custom, or habit.

The materials mentioned above help students copyedit their work and improve their writing. Production editing is another matter. Perhaps the best work on production editing is Ken Caird's *Cameraready*. It covers a wealth of facts on moving a manuscript from typed form to final printed form. The problem is its date: teachers will have to supplement from other sources. Another, *Pasteups and Mechanicals* by Jerry Demoney and Susan Meyer, focuses on getting text into shape for the printer. Kenneth Hird's *Paste-up for Graphic Arts Productions* covers much of the same material. Two recent books, *Getting*

It Printed by Mark Beach, Steve Shepro, and Ken Russon and *Communication Graphics* by Wendell C. Crow, provide students with excellent suggestions on moving from camera-ready to printed form while maintaining schedules and budgets.

Two older books, *Words into Type* (under revision) and Edmund Arnold's *Ink on Paper 2*, do not consider contemporary innovations such as desktop publishing (where the text processor and the formatter are in one program), but they (especially *Words into Type*) do have information the future professional technical writer needs to understand about how materials move from manuscript to the printed page.

The problem of desktop publishing (particularly in relation to mainframe publishing) offers special difficulties for the teacher of an advanced course. The main sources of information are articles and specialized journals. Because desktop publishing is so dependent on hardware (the programs for the Apple do not work on the IBM or DEC Rainbow), no single source is really adequate. *Publish!* attempts to cover as many programs and systems as possible. Hardware-specific magazines, such as *Hardcopy* and the *DEC Professional*, likewise provide occasional articles. Certainly there is enough material around for students to become familiar with the topic.

Mainframe publishing (where there are separate programs for text processing and formatting) offers even more of a challenge. Magazines such as *Unix Review* offer the occasional article on making effective use of formatting programs. The best approach is to ask the computer center for information on such programs as *Script* (and Robert Seidel and Charles W. Gainey's *Script/PC*) and public-domain formatters (such as J. Anthony Movshon's *FPRINT* and Carl D. Neiburger's *TypeStyl*).

The central problem facing the teacher of advanced technical writing is what to teach. As I have tried to make clear in this brief space, more of the same from an intermediate-level class is not beneficial to the student who wants to be a professional technical writer. Instead, the student needs information on the kinds of writing tasks demanded by business, industry, or government. This essay has provided a few suggestions for materials the teacher can use. Materials from current journals, from anthologies, from the Society for Technical Communication, and from the places of employment will provide additional information the student needs.[1]

NOTE

[1] In addition to the works cited in the essay, resources include the annual bibliography in *Technical Writing Teacher*; anthologies from the Society for Technical Communication and Teachers of Technical Writing; the proceedings from various societies, such as the Society for Technical Communication, Council for Programs in Technical and Scientific Communication, and Society for Scholarly Publishing; the section "Recent and Relevant" in each issue of *Technical Communication*; and the following books: Alred, Reep, and Limaye, *Business and Technical Writing: An Annotated Bibliography of Books, 1880–1980*; W. S. Anderson and Cox, *The Technical Reader*; Bacon et al., *Writing*

Winning Proposals; Behling, *Guidelines for Preparing the Research Proposal*; Bowen and Mazzeo, *Writing about Science*; Brand and White, *Legal Writing*; Carlson et al., *An Annotated Bibliography on Technical Writing, Editing, Graphics, Publishing, 1966–1980*; Felker, *Document Design*; Journet and Kling, *Readings for Technical Writers*; Larson, *How to Write a Winning Proposal*; Leonard and McGuire, *Readings in Technical Communication*; Lynch and Swanzey, *The Example of Science*; Odell and Goswami, eds., *Writing in Nonacademic Settings*; Poter-Roth, *Proposal Development*; Schatzberg, Waite, and Johnson, *The Relations of Literature and Science: An Annotated Bibliography of Scholarship, 1880–1980*; Whalen, *Writing and Managing Winning Technical Proposals*.

The High-Tech Workplace: Implications for Technical Communication Instruction

ELIZABETH TEBEAUX

SINCE 1981, BUSINESS, trade, and professional publications have focused sharply on trends resulting from the impact of computer technology on the workplace. For example, consider the following: Between 1860 and 1980, the proportion of American workers involved in agriculture and manufacturing declined from 83% to 30%. As advanced technology is applied to agriculture and heavy industry during the last decade of this century, the proportion of the work force employed in manufacturing is expected to decline gradually to about 3% (Best 63; Cook, "You Mean" 143). In contrast, since 1950, the number of jobs in service industries has risen by nearly 146% (Ehrenhalt 16). Today, 55% of the workers in the United States are employed in information industries. More people are involved in information-communication services than in mining, agriculture, manufacturing, and personal services combined. By 2000, 80% of the work force will be employed in jobs that involve either generating or transmitting information (Feingold 10; Cetron, Rocha, and Luckins 30). In 1980, there was one electronic work station for every twenty-three white-collar employees; by 1989, one for every two.

This growth in information technology portends enormous changes in the way managers obtain information, make decisions, and perform their work (Gray 69). With data processing, telecommunications, and administrative services now merged, the $200 billion electronics industry plays the central role in the United States and the world economies that automobiles played in the past. Communication equipment and services are expected to contribute about 40% of world industrial value—or about $1.5 trillion per year—by the turn of the century (International Data, "Revolution"; A. Smith). The Information Age, not the Industrial Age, now defines the world of work (Drucker). The replacement of brawn industries with brain industries has changed both the jobs available to college students and the competencies necessary to perform these jobs. Because the goal of technical communication courses is to prepare students to communicate successfully at work, technical communication teachers need to understand how the emergence of the Information Age is affecting the work environment and how changes in work will alter what we should teach.

Information Technology and the Redefinition of Work

Steady advances in computer technology and information systems are making many jobs obsolete and drastically reshaping others. One result is large-scale dislocation of employees, many of whom are undereducated for high-technology jobs and have no place to go (Ehrbar; Banks; Horvath). Despite the addition of over 10,000 new job titles to the Department of Labor job list, many jobs our students will hold a decade from now do not yet exist. Students must be prepared for the possibility of working in areas totally unrelated to their academic fields, and they should know that formal academic training for these coming jobs will not always exist (B. Jones 181; Cook, "The Knack"; Mattill; Salino; Schlefer).

To remain employed in an unpredictable job market, students can no longer depend on the future relevance of today's technical knowledge, which is becoming outmoded by the growth in knowledge stemming from information technology. According to one study,

> The half-life of an engineer's knowledge today is five years. In 10 years, 90 percent of what an engineer knows will be computer-related. Eighty-five percent of the information on National Institutes of Health computers is upgraded every five years. The rapidly changing job market, along with the changing requirments of new technologies, will necessitate increased training across the board. Up to 4 percent of the labor force will be in job retraining programs by the 1990s. Because of fundamental changes in the economy, there will be fewer and fewer well-paying jobs not requiring advanced training. . . . Schools will be used to train both children and adults. The academic day will be lengthened to seven hours for children; adults will work a 32-hour workweek and prepare for their next job in the remaining hours. (Cetron, Rocha, and Luckins 33–34)

Formal education will remain crucial ("The Growing Need"). Education that prepares students for the uncertain decades ahead will emphasize analytical skills, learning strategies, and concepts rather than technical detail. As Samuel Ehrenhalt of the Department of Labor has predicted, "If anything, the character of work is continuing to become less routine, more problem-solving, and above all, more changeable. That requires a mobile, flexible, adaptable labor force. The ability to learn will, in my view, emerge as the premium skill of the future. It is not what you know that will be the key to success in the emerging economy, but what you can learn—and how fast" (15–16). Managers, technicians, and engineer-scientists must be able to live and produce in an environment where holistic and relational thinking is routine (Rehder and Porter 53; M. Thomas; Kotter). Because work will become increasingly computer-mediated, tasks will be accomplished through information systems rather than through direct physical contact with the object of the task. Imagination, analysis, and the ability to create theoretical insight, rather than hands-on experience, will become essential competencies for developing concepts to handle the growing masses of available information (Zuboff; United States Congress). Because what one learns today will quickly become obsolete, staying employed will require commitment to a lifetime of learning.

In short, technology has changed not only work and the knowledge required to perform work but also national and international economics, demographics, and the structures of society. As studies in the entire issue of the Fall 1987 *Occupational Outlook Quarterly* indicate, employees can anticipate several career changes during their lifetimes. As jobs and careers become obsolete, they will be replaced by new kinds of work requiring new kinds of training and relocation to a constantly changing array of work sites, many beyond United States boundaries (Kutscher; White).

Information Technology and the Redefinition of Communication Skills

From a communication perspective, the massive generation and dissemination of data and information from knowledge industries mean that communication skills will continue to be essential for a successful career and that the demand will grow to help others plan, manage, evaluate, and disseminate information to a wide range of users, many of whom will have inadequate literacy skills. Communication skills will be redefined as the ability to handle vast amounts of information and to adapt it for a wide range of users: the foremost problem that will confront employees in the information society emanates from the quantity of information being continuously generated. Because the amount of information generated doubles every two years or less, the central question becomes how to handle it—access it, manage it, use it, and help others use it.

From an organizational and a communication standpoint, a major concern is that information is not being managed. "It is available in overabundance or not at all. Seldom timely or complete, it is provided at a cost that cannot be determined and is given almost totally subjective value" (Connell 29). Information accumulates in additive, digital lists, while meaning, being subjective and referring to synthetic or holistic properties that cannot be reduced to the sum of parts, does not come easily or inevitably from a growing heap of mere information (Klapp). The problem during the next century lies not in the amount of scientific information available but in how to recognize and gain access to the value and significance of a large body of data. Such a plethora of technical options has been made available that society is threatening to fall behind in the orderly processing of information (Horowitz 630).

Managing information overload will require employees who can effectively deal with people, organizational problems, oral and written communication, economic and financial analysis, and computer-information systems—in short, employees with a broad education rather than with a specialized training in a narrow field (Gray; Kiplinger 42; Jackson; Hull and Pedrotti). Communication skills will focus on successful extraction and adaptation—on knowing how to sift, select, shape, and present information to users as well as how and when to use graphics and oral media as replacements for written communication. Clarity and conciseness will also continue to be important qualities, as employees will be inundated with information that they need to understand and apply quickly and effectively; minimizing paperwork will also be a growing trend (Lampe; Denise; Pittel). Because communication skills will be essential in making information useful, analysis of audience (users) will become even more crucial to meet new, more demanding communication situations than those now confronting us.

Information Technology and Organizational Communication

Communication technology is also reshaping organizational hierarchies and the way decisions are made and communicated in organizations. Because rapid changes in technology require rapid decisions, rigid, multilevel hierarchies in organizations are being replaced with less complex decision-making structures requiring fewer levels. Z theory has made participatory management more widespread, and networks—composed of individuals from several areas of the organization—are being used to solve problems. These decision-making teams merge representatives from traditionally separate business functions (design, engineering, production, purchasing, distribution, marketing, sales, e.g.) into a problem-solving body. An employee's effectiveness in this kind of decision-making group will depend on communicating clearly, understanding the perceptions of others, and knowing how to work comfortably with individuals with diverse technical and organizational backgrounds. No hierarchy controls this kind of problem solving; solutions may come from anyone. The collective capacity of the group becomes something more than the simple sum of its members' skills; paperwork is kept to a minimum; and the network itself dissolves after the problem is solved. Employees may find themselves working in several networks at once, each brought into existence to solve an immediate organizational problem. Rigid, extensive communication hierarchies in organizations, such as those described by Mathes and Stevenson in 1976, are diminishing (Reich; Fraker; Brightman and Verhoeven; McInnis; Hine; Grayson, "Networking").

As teleconferencing, computer conferencing, and electronic mail expand, writers often "communicate" only by computer with individuals or groups of individuals unknown to the writer. When communication becomes only computer-mediated and when face-to-face communication is not possible, behavior in the communication process changes. Research has already shown that great care needs to go into the design of computer messages, with emphasis on the following points: building sentences that are semantically and syntactically unambiguous, developing messages that are concise, and avoiding a tone that is overbearing in trying to reach the audience beyond the computer. Communication strategies change dramatically when common dramaturgical cues —shared psychological space, social space to produce cues necessary to create a shared experience, facial expression, tones of voice, and personal interaction with the audience—are missing (Sproull and Kiesler; Kiesler, Siegel, and McGuire; Wolff, "What You Should Know"; Grayson, "Productivity"; Magee and Little). Particularly in computer conferencing, writers may help build files of information that may be accessed by an unknown number of readers. In such a context, writers also have to be careful to provide clear background, rationale, and purpose so that readers can accurately and quickly perceive the significance and meaning of the material presented.

Information Technology and the Growth of International Communication

The high-tech information society has also merged telecommunications, data processing, and information management to produce worldwide communication systems (Walt). This reality means that an increasing number of United States industries will proliferate into multinational companies. The globe is rapidly becoming a single marketplace where goods are being made wherever they can be made the cheapest, regardless of national boundaries (Kiplinger 32–34; Bolt). The term "foreign affairs" is thus rapidly becoming an anachronism, and United States employees will need to know how to communicate with employees of different cultures and understand the culturally accepted prerequisites of communicating and working effectively with employees in other cultures (Moore; Ravitch; "Two-Hatted M.B.A.").

Planning Technical Communication Research for the Information Age

Teaching technical communication effectively in the constantly changing Information Age is not going to be easy. Textbooks are not adequate and probably never will be because of the lag between writing and publication. But new approaches in texts need to be attempted that reflect the broad competencies required for communicating in the high-tech workplace. Teaching generic reports and letters does not prepare students for the nongeneric kinds of writing now done routinely in high-tech organizations. Communication models do not reflect problems of information overload faced by employees who have too much material to read and process. Cognitive researchers are only beginning to deal with decoding problems confronting receivers faced with heaps of raw data from which "meaning" must be extracted and then transmitted. Research results about effective use of voice, syntax, and graphics are far from consensual. Research in intercultural communication, human-computer interface, and the relations among data, information, and communication is still germinal. As Jim Corder stated so convincingly, we need a new rhetoric, and "If We Do Get There, There Won't Be There Anymore" (161). The new rhetoric must be one that deals with all kinds of discourse, visual messages, electronic communications, mass media, and the nature of knowledge (how we know what we know). Research in all these areas becomes imperative if we are to know and understand what we should teach.

In the meantime, we need to reexamine current communication models, survey trends in design of reports for electronic transmittal, learn protocols of communicating in other cultures, stay abreast of efforts to access data bases and information systems more efficiently, and continue research to determine how computer composing modifies the composing process. Never in the history of technical communication has practical, empirical research been so crucial to effective instruction.

Planning Instructional Changes for the Information Age

The inextricable role of communication in both the work environment and the definition of work itself suggests that professional communication courses will remain crucial. Changes in the work environment, moreover, have important implications for how we approach and teach technical communication, particularly on the advanced level, to ensure its relevance to students in the years ahead.

First, approaches to technical communication must reflect, more than texts currently do, the new constraints imposed by technology on the workplace. Specifically, the approach to technical communication must stress how communication in the high-tech workplace differs from communication in academic contexts. Advanced courses in technical communication can no longer be advanced skills courses. Instead, the focus should be on understanding the Information Age and on giving students practice in developing the competencies necessary for becoming effective communicators.

Second, for a number of reasons, we need to stop segregating our technical communication students according to their academic majors. On the job, students will collaborate with employees from diverse academic backgrounds. Many students will work in jobs unrelated to their academic fields of study. Studies of employee communication needs do not justify creating separate communication courses for students with different academic majors (Anderson, "What

Survey Research"). The increasing use of networks and other forms of small-group decision making suggests that students will benefit from learning to communicate with other students from diverse academic backgrounds. This kind of interdisciplinary communication environment, for which we should be preparing students, is clearly described by Thomas Peters and Robert Waterman throughout *In Search of Excellence* and by Peters and Nancy Austin in *A Passion for Excellence*. Business studies continue to emphasize that one solution to productivity and quality problems resides in breaking down barriers between technical culture and business culture and using continuing, interactive, dynamic planning and problem solving among areas within organizations (Cannon; Westwood).

Realigning courses to include students from a range of disciplines provides a laboratory setting where students can practice network problem solving that requires more than writing technical solutions to discipline-specific problems. In refocusing our courses to include networking, we can have teams of students from a variety of disciplines analyze and then recommend, individually and as a group, solutions to cases reflecting the kinds of organizational problems they can expect to face on the job. Group problem solving can also show students how oral and written communication relate and how written solutions often emanate from oral analysis.

Third, in response to both networking and the growing quantity of information, technical communication courses must intensify the study of audiences (as users of information), of the purposes for which information may be used, and of the contexts in which information may be processed and then applied. Writing assignments must help students understand how to generate reports from data, that is, how to extract information from quantities of raw data, analyze it, and sift, shape, and summarize it to enable readers to make efficient, accurate decisions. Students need to learn the difference between operational data and management data and understand how to highlight the significance of each type for the needs of a particular level of user.

Fourth, the massive training and retraining that provide displaced employees with new skills indicate that managers will need to know how to develop work procedures and instructions. Instructions and procedures should be emphasized in advanced professional-writing courses, but courses in designing and preparing training, operations, and procedure manuals should be added. Because computers are becoming the main means of providing continuing education, these courses should also cover how instructional materials can be adapted to computer presentation (Kendel and Benoir). Because many organizations are now developing their own business software or modifying existing software, courses in the development of effective user documentation should be made available on the graduate and undergraduate levels.

Fifth, computers have antiquated our traditional approach to teaching graphics. Instead of continuing to emphasize conventions for developing graphics and choosing the best graphic for a given kind of information, we need to develop case problems that require students to use computer graphics and data-analysis packages to analyze and present information. Cases should be based on raw data that must be made meaningful to the intended users. The design of these cases needs to make two points clear: (1) in some cases, visual display is more effective than verbal presentation, and (2) writers will often have to decide when to use visual display and when to use verbal presentation.

Sixth, the growing quantity of information now available in every field has affected how we should teach research. While students need to learn how to use abstracts, indexes, and specialized reference materials in their fields, this approach by itself does not enable students to perceive the access problem caused by burgeoning data. We must also emphasize how current search systems are devised and organized, the kinds of systems available, theories of indexing, the changes that standard research systems (like card catalogs and major indexes) are undergoing, and procedures for developing search strategies to access online material. The stark fact is that no common computer protocol or language exists to synthesize the two or three hundred data bases currently available (Horowitz).

Seventh, because of emerging worldwide communication networks, students need to understand that communications they routinely generate on the job may be accessed and used by employees in other cultures who have access to global electronic mail and computer conferencing. Advanced courses and programs need to include instruction in intercultural communication: how to deal with cultural barriers and differences in planning and writing for multinational audiences. We also need to stress to students the value of a broad liberal arts education—for example, mastering another language and understanding other cultures and the political and religious systems that underlie intercultural communication barriers and existing protocols (Kiplinger 42; Wolff, "When Lab"; Steele; Maisonrouge).

But perhaps the most important change involves helping students understand the proper relation between computer literacy and human analysis and perception. Our instruction needs to make clear that the burgeoning growth of information has transformed the communication process but that giving readers information does not mean that communication has occurred and that computer literacy and word processing are not substitutes for knowing how to communicate or for understanding the process for developing communication. Students need to understand that quantities of information and advances in information systems do not inevitably make us better informed or inevitably enhance our comprehension of information (Wessel; Dreyfus and Dreyfus). As An Wang has stated, our entire thought and meaning process is impervious to automation (International Data, "Office Systems"). Computer and word processing skills are no substitute for the ability to analyze, to create the shared-language experience necessary for writer and reader to communicate. Word processing often disguises noncommunicating information clothed in grammatical correctness (Parret).

Ultimately, the increasing speed by which information is generated and documents are processed makes even more important the need to stress the main principle that underpins every professional writing course: "Communication occurs when people are led to experience shared perceptions and assumptions about what is real, what is relevant, and what is important in a particular situation" (Sanderlin 42). Our students must understand that sifting through and analyzing the growing pool of information on every subject is only the first rite of passage to the more demanding exercise—the painful intellectual deliberations enabling us to help readers gain understanding. Information itself does not dispell ignorance. Knowledge does (Landvater; Menofsky; M. Nelson).

Notes on the Contributors

Jo Allen is assistant professor of English and director of the Writing Center at East Carolina University. She teaches undergraduate and graduate courses in literature and technical communication and sponsors the Society for Technical Communication Student Chapter, which she helped found. Coauthor of *Teaching Technical Writing in the Secondary School*, she has written articles that have appeared in *Technical Communication, Teaching English in the Two-Year College*, and numerous proceedings. She is currently conducting research on rhetorical schemes in classical technical and scientific literature.

Mary Beth Debs is assistant professor of English at the University of Cincinnati and director of the Writing Program, which offers undergraduate and graduate courses in technical writing, journalism, and creative writing. In 1984 the Department of English Writing Program received an Academic Challenge Award from the Ohio Board of Regents. Her research interests include institutional rhetoric, collaborative writing, and technical communication. Her work has been published in the *Journal of Business and Technical Communication*, the *Technical Writing Teacher*, and the *Journal of Business Communication*.

Bertie E. Fearing is professor of English, codirector of Technical Communication Programs, and assistant chair of the department of English at East Carolina University, where she also directs the Chancellor's Forum. She is editor emeritus of *Teaching English in the Two-Year College*, past associate editor of *Technical Communication*, and an editorial board member of the *Journal of Advanced Composition*. Her publications appear in *Research in Technical Communication: A Bibliographic Sourcebook, Writing Centers: Theory and Administration, Technical and Business Communication in Two-Year Programs*, and several journals. She has served on the executive committees of the Conference on College Composition and Communication and the National Council of Teachers of English.

Roger A. Grice is adjunct professor of technical communication at Rensselaer Polytechnic Institute and an advisory information developer at the IBM laboratory in Kingston, New York. He has been a member of IBM's information-development organization for twenty-four years and is currently involved in large-systems documentation, information usability, and online information. He is a senior member of the Society for Technical Communication and manager of its scholarships and membership committees. He is also a senior member of the Institute of Electrical and Electronics Engineers (IEEE). His work is included in *Text, Context, and Hypertext* and in numerous journals and proceedings.

Michael P. Jordan is associate professor of linguistics and technical communication at Queen's University, Kingston, Canada; he holds an honorary fellowship in English language at the Hatfield Polytechnic, England. His research activities began some twenty years ago, when as an engineer he became dissatisfied with the traditional teaching of technical writing. The results of his work are included in *Rhetoric of Everyday English Texts* and *Fundamentals of Technical Description* as well as in numerous articles, for which he has received several awards. He is now the president of the Canadian Association of Teachers of Technical Writing.

MARY M. LAY is associate professor and chair of Clarkson University's Department of Technical Communications. She teaches advertising copy and design as well as technical writing. She is author of *Strategies for Technical Writing: A Rhetoric with Readings* and coeditor of the forthcoming Baywood Technical Communication Series text *Collaborative Writing in Industry*. Her work has appeared in *Technical Communication*, the *Technical Writing Teacher*, *IEEE Transactions on Professional Communication*, the *Journal of Technical Writing and Communication*, and other journals. She is the 1988–90 president of the Association of Teachers of Technical Writing.

CAROLYN R. MILLER is associate professor of English at North Carolina State University. Her research interests are in rhetorical theory and the role of rhetoric in science and technology. She is a contributor to *Research in Technical Communication: A Bibliographic Sourcebook* and *Writing in Nonacademic Settings* and coeditor of *New Essays in Technical Communication: Research, Theory, Practice*.

JOHN H. MITCHELL is a retired professor of English at the University of Massachusetts, where he taught for thirty-five years. He has been visiting professor at Arizona State University, University of Hawaii, Canberra College of Advanced Education, and University of Victoria. He is author of two books, *Handbook of Technical Communication* and *Writing for Technical and Professional Journals*, and many professional articles. In recognition of his notable contributions to scientific and technical communication, he has been made a fellow of five major organizations, including the Society for Technical Communication, the Association of Teachers of Technical Writing, and the Institute for Scientific and Technical Communication.

JANICE C. REDISH is a vice president of the American Institutes for Research in Washington, D.C., and director of AIR's Document Design Center. A linguist by training, she devotes her research to the process of writing, writing in organizations, and making information accessible to users. Her major publications include *Guidelines for Document Designers* and *Writing in the Professions* as well as chapters in *Writing in Nonacademic Settings, Literacy for Life*, and *The English Language Today*. She is a member of the executive board of the *Journal of Technical Writing and Communication*.

DAVID A. SCHELL is a usability specialist for IBM. When he wrote his contribution for this volume, he was acting director of the Usability Test Laboratory of the American Institutes for Research in Washington, D.C., where he developed AIR's logging program. A cognitive psychologist, he conducts research on human memory and reading and how people acquire and use knowledge. He is also a skilled programmer who conducts online experiments.

JACK SELZER is associate professor of English at Pennsylvania State University, where he directs the department's composition programs and teaches a variety of undergraduate and graduate courses in composition, technical and business writing, and literature. His work on rhetoric and literature has appeared in numerous books and journals, including *College Composition and Communication, Writing in Nonacademic Settings, Journal of Business Communication, Philological Quarterly, English Literary Renaissance,* and *Journal of Technical Writing and Communication.*

MARION K. SMITH is associate professor of English at Brigham Young University, where he teaches American literature, science fiction, and technical writing. He came to academe after extensive vocational experience in farming, ranching, construction, electronics, and small-business management. He has thirty-five years of active and reserve duty with the United States Army and Air Force and is currently with the Tactical Air Command. His publications include *Technical Writing, Formal Report Format,* journal articles, and conference proceedings papers.

JAMES W. SOUTHER, professor emeritus of Scientific and Technical Communication at the University of Washington, is a major contributor to our field. His *Technical Report Writing* (with M. L. White) was the first process-approach textbook; and his Westinghouse management study, our first significant research project on audience analysis. His contributions and leadership have led to his being selected a fellow in both the Society for Technical Communication and the Association of Teachers of Technical Writing and to his receiving the Lifetime Achievement Award from the Professional Communications Society of the IEEE.

W. KEATS SPARROW is professor of English, codirector of Technical Communication Programs, and chair of the department of English at East Carolina University. He is a founding editor of *Teaching English in the Two-Year College,* a frequent contributor to journals, coeditor of *The Practical Craft: Readings for Business and Technical Writers* and *Technical and Business Communication in Two-Year Programs,* and coauthor of *Business Communications.* His earlier essay collection, *The Practical Craft: Readings for Business and Technical Writers,* won the Best Collection of Essays Award from the National Council of Teachers of English Publication Awards Program in Technical and Scientific Communication. He has served on the NCTE's Committee on Technical and Scientific Communication.

ELIZABETH TEBEAUX is associate professor of English and coordinator of technical writing at Texas A&M University, where she also develops information for the Texas A&M system. She chairs the CCCC Conference on Technical and Scientific Communication and is member-at-large of the Association of Teachers of Technical Writing. She has written a number of papers on the pedagogy of technical communication and on style in literature and technical communication. Her work appears in *College Composition and Communication, College English, Journal of Technical Writing and Communication,* and *Journal of Advanced Composition.* She is coauthor of *Written Communication in Business and Industry.*

ISABELLE KRAMER THOMPSON is associate professor of English at Auburn University, where she teaches in the Professional Communication Program. Her articles applying research in linguistics and rhetoric to the teaching of writing have appeared in the *Journal of Technical Writing and Communication, Rhetoric Review, Teaching English in the Two-Year College,* and *Technical Communication.* She is the coauthor of *Communication for Technicians: Reading, Writing, and Speaking on the Job.*

THOMAS L. WARREN is professor of English, director of the Technical Writing Program, and coordinator of Graduate Programs in English at Oklahoma State University. He is a past president of the Council of Programs in Scientific and Technical Communication, a fellow of the Society for Technical Communication, and a fellow of the Association of Teachers of Technical Writing. In addition to his many articles, he is author of *Technical Communication: An Outline, How to Write Technical Reports,* and *Technical Writing: Purpose, Process, and Form.*

Works Consulted

Aldefer, Clayton, and Ken Smith. "Studying Intergroup Relations Embedded in Organizations." *Administrative Science Quarterly* 27 (1982): 35-65.

Allen, J. W. "Introducing Invention to Technical Students." *Technical Writing Teacher* 5 (1978): 45-49.

Allen, Nancy, et al. "What Experienced Collaborators Say about Collaborative Writing." *Iowa State Journal of Business and Technical Communication*, forthcoming.

Alred, Gerald J., Diana C. Reep, and Mohan R. Limaye. *Business and Technical Writing: An Annotated Bibliography of Books, 1880–1980*. Metuchen: Scarecrow, 1981.

Anderson, John R., and Brian H. Ross. "Evidence against a Semantic-Episodic Distinction." *Journal of Experimental Psychology: Human Learning and Memory* 6 (1980): 441-65.

Anderson, Paul V. *The Architecture of Cognition*. Cambridge: Harvard UP, 1983.

———. Introduction. *Education*. Spec. issue of *Technical Communication* 31.4 (1984): 4-8.

———. "Survey Methodology." Odell and Goswami, *Writing* 453-501.

———. *Technical Writing*. San Diego: Harcourt, 1987.

———. "What Survey Research Tells about Writing at Work." Odell and Goswami, *Writing* 3-83.

———. "What Technical and Scientific Communicators Do: A Comprehensive Model for Developing Academic Programs." *IEEE Transactions on Professional Communication* PC-27.4 (1984): 161-67.

Anderson, Paul V., R. John Brockmann, and Carolyn R. Miller, eds. *New Essays in Technical and Scientific Communication: Research, Theory, Practice*. Farmingdale: Baywood, 1983.

Anderson, W. Steve, and Don Richard Cox, eds. *The Technical Reader: Readings in Technical, Business, and Scientific Communication*. New York: Holt, 1984.

Andrews, Deborah C., and Margaret D. Blickle. *Technical Writing: Principles and Forms*. 2nd ed. New York: Macmillan, 1982.

Aristotle. *Nicomachean Ethics*. Trans. J. A. K. Thompson. New York: Penguin, 1955.

———. *Rhetoric*. Trans. Lane Cooper. Englewood Cliffs: Prentice, 1932.

Arms, Valarie. "Collaborative Writing on a Word Processor." Conf. Record of the IEEE Professional Communication Soc. 19-21 Oct. 1983. New York: Inst. of Electrical and Electronics Engineers, 1983. 85-86.

Arnold, Edmund C. *Ink on Paper 2: A Handbook of the Graphic Arts*. New York: Harper, 1972.

Atlas, M. A. "The User Edit: Making Manuals Easier to Use." *IEEE Transactions on Professional Communication* PC-24.1 (1981): 28-29.

Bacon, Terry R., et al. *Writing Winning Proposals*. Bontiful: Shipley, 1987.

Baldwin, R. Scott, et al. "Effects of Topic Interest and Prior Knowledge on Reading Comprehension." *Reading Research Quarterly* 20 (1985): 497-504.

Banks, Howard. "General Electric: Going with the Winners?" *Forbes* 26 Mar. 1984: 97-106.

Barnett, George A., and Carol Hughes. "Communication Theory and Technical Communication." Moran and Journet 39-83.

Bazerman, Charles. "Codifying the Social Scientific Style: The *APA Publication Manual* as a Behaviorist Rhetoric." Iowa Symposium on the Rhetoric of the Human Sciences. Iowa City, 1984.

———. "Modern Evolution of the Experimental Report in Physics: Spectroscope Articles in *Physical Review*, 1893–1980." *Social Studies of Science* 14 (1984): 163-96.

———. "Physicists Reading Physics: Schema-Laden Purposes and Purpose-Laden Schema." *Written Communication* 2 (1985): 3-23.

———. "Scientific Writing as a Social Act: A Review of the Literature of the Sociology of Science." Anderson, Brockmann, and Miller 156-84.

———. "The Writing of Scientific Non-fiction: Contexts, Choices, Constraints." *PrelText* 5 (1984): 39-74.

Beach, Mark, Steve Shepro, and Ken Russon. *Getting It Printed: How to Work with Printers and Graphic Services to Assure Quality, Stay on Schedule, and Control Costs.* Portland, OR: Coast to Coast, 1986.

Beekman, John, and John Callow. *Translating the Word of God.* Grand Rapids: Zondervan, 1974.

Begg, Ian, et al. "On Believing What We Remember." *Canadian Journal of Behavioral Science* 85 (1985): 621-34.

Behling, John H. *Guidelines for Preparing the Research Proposal.* Rev. ed. Lanham: UP of America, 1984.

Belanger, Joe, and Glenn R. Martin. "The Influence of Improved Reading Skill on Writing Skill." *Alberta Journal of Educational Research* 30 (1984): 194-212.

Benson, Philippa J. "Writing Visually: Design Considerations in Technical Publications." *Technical Communication* 32.4 (1985): 35-39.

Berkenkotter, Carol. "Decisions and Revisions: The Planning Strategies of a Publishing Writer." *College Composition and Communication* 34 (1983): 156-72.

Bernstein, Richard J. *Praxis and Action: Contemporary Philosophies of Human Activity.* Philadelphia: U of Pennsylvania P, 1971.

Best, Fred. "Work in a High-Tech Future." *Futurist* Apr. 1984: 63.

Bizzell, Patricia. "Cognition, Convention, and Certainty: What We Need to Know about Writing." *PrelText* 3 (1982): 213-43.

———. "Composing Processes: An Overview." *The Teaching of Writing.* Ed. David Bartholomae. 85th Yearbook of the Natl. Soc. for the Study of Education. Chicago: U of Chicago P, 1986. 49-70.

Blanchard, Harry E. "A Comparison of Some Processing Time Measures Based on Eye Movement." *Acta Psychologica* 58 (1985): 1-15.

Blau, Judith. "Patterns of Communication among Theoretical High Energy Physicists." *Sociometry* 37 (1974): 391-406.

Blickle, Margaret D., and Kenneth W. Houp. *Reports for Science and Industry.* New York: Holt, 1958.

Blicq, Ron S. *Technically Write! Communicating in a Technological Era.* 2nd ed. Englewood Cliffs: Prentice, 1981.

Bloom, Harold. *The Anxiety of Influence: A Theory of Poetry.* London: Oxford UP, 1973.

Bolt, James F. "Global Competitors: Some Criteria for Success." *Business Horizons* Jan.-Feb. 1988: 34-41.

Bowen, Mary Elizabeth, and Joseph A. Mazzeo, eds. *Writing about Science*. New York: Oxford UP, 1979.

Bradford, Annette Norris. "Conceptual Differences between the Display Screen and the Printed Page." *Technical Communication* 31.3 (1984): 13-16.

Brand, Norman, and John A. White. *Legal Writing: The Strategy of Persuasion*. 2nd ed. New York: St. Martin's, 1988.

Brightman, Harvey J., and Penny Verhoeven. "Running Successful Problem-Solving Groups." *Business* Apr.-June 1986: 15-23.

———. "Why Managerial Problem-Solving Groups Fail." *Business* Jan.-Mar. 1986: 24-29.

Broadhead, Glenn, and Richard Freed. *The Variables of Composition*. Carbondale: Southern Illinois UP, 1986.

Brockmann, R. John. "Advisory Boards in Technical Communication Programs and Classes." *Technical Writing Teacher* 9 (1982): 137-46.

Brown, W. C., and Robert E. Tuttle. *Writing Useful Reports: Principles and Applications*. New York: Appleton, 1956.

Brown, Zoe W. "Page Design for Data Processing User Documentation." Proc. of the Intl. Technical Communication Conf. 29 Apr.–2 May 1984. Washington: Soc. for Technical Communication, 1984. VC-27-30.

Browning, Christine. *Guide to Effective Software Technical Writing*. Englewood Cliffs: Prentice, 1984.

Bruffee, Kenneth. "Collaborative Learning: Some Practical Models." *College English* 35 (1973): 634-42.

Brushaw, Charles T., Gerald J. Alred, and Walter E. Oliu. *Handbook of Technical Writing*. 3rd ed. New York: St. Martin's, 1987.

Burke, Kenneth. *A Grammar of Motives* and *A Rhetoric of Motives*. New York: World, 1962.

Button, Jim. *PC-Style*. Computer software. Buttonware, 1986.

Caird, Ken. *Cameraready*. San Diego: Univelt, 1973.

Cannon, Peter. "Integrative Planning and Communication of Research." *Research Management* May-June 1984: 20-23.

Carlson, Helen V., et al. *An Annotated Bibliography on Technical Writing, Editing, Graphics, and Publishing, 1966-1980*. Washington: Soc. for Technical Communication, 1983.

Carroll, J. M., and R. L. Mack. "Actively Learning to Use a Word Processor." *Cognitive Aspects of Skilled Typewriting*. Ed. W. Cooper. New York: Springer, 1983. 259-82.

Castleberry, K. Sue. "Comprehension, Metacomprehension, and Instructional Implications for College Students." *Reading World* 23 (1984): 204-08.

CBE Style Manual Committee. *CBE Style Manual*. 5th ed. Bethesda: Council of Biology Editors, 1983.

Cetron, Marvin J., Wanda Rocha, and Rebecca Luckins. "Into the 21st Century: Long-Term Trends Affecting the United States." *Futurist* July-Aug. 1988: 29-40.

Chandani, Ashok. "Writing Styles of Abstracts in Occupational Therapy Journals." *British Journal of Occupational Therapy* 48 (1985): 244-46.

Charland, Dennis A. "Online Documentation: Promise and Problems." Proc. of the Intl. Technical Communication Conf. 19 Apr.–2 May 1984. Washington: Soc. for Technical Communication, 1984. WE-158-61.

Charney, Davida, and Lynn M. Reder. "The Role of Examples and Explanations in Teaching Procedural Skills." Conf. on College Composition and Communication. Minneapolis, Mar. 1985.

Charrow, V. R. "Language in the Bureaucracy." *Linguistics and the Professions.* Ed. R. J. Di Pietro. Norwood: Ablex, 1982. 173-88.

Chavarria, Linda Stout. "More Online Computer Documentation in the Future—Are You Ready?" Proc. of the Intl. Technical Communication Conf. 1–4 May 1983. Washington: Soc. for Technical Communication, 1983. ATA-22-25.

Clarke, Beverly. "Multiple Authorship Trends in Scientific Papers." *Science* 143 (1964): 822-24.

Clifford, John. "Composing in Stages: The Effects of a Collaborative Pedagogy." *Research in the Teaching of English* 15 (1981): 37-53.

Clough, M. Evalyn, and Thomas J. Galvin. "Educating Special Librarians: Toward a Meaningful Practitioner-Educator Dialogue." *Special Libraries* 75 (1984): 1-8.

Cocklin, Thomas G., et al. "Factors Influencing Readability of Rapidly Presented Text Segments." *Memory and Cognition* 12 (1984): 431-42.

Coggin, Bill. "Better Educational Programs for Students of Technical Communication." *Technical Communication* 27.2 (1980): 13-17.

Cole, Jonathan. "Patterns of Intellectual Influence in Scientific Research." *Sociology of Education* 43 (1970): 377-403.

Commission on Graduate Studies in Public Relations. *Advancing Public Relations Education: Recommended Curriculum for Graduate Public Relations Education.* New York: Foundation for Public Relations Research and Education, 1985.

Coney, Mary B., and James W. Souther. "Analytical Writing Revisited: An Old Cure for a Worsening Problem." *Technical Communication* 31.1 (1984): 4-8.

Connell, John J. "The Future Office: New Techniques, New Career Paths." *Personnel* July-Aug. 1983: 23-32.

Connors, Robert J. "The Rise of Technical Writing Instruction in America." *Journal of Technical Writing and Communication* 12 (1982): 329-52.

Cook, Claire Kehrwald. *Line by Line: How to Edit Your Own Writing.* Boston: Houghton, 1985.

Cook, James. "The Knack . . . and How to Get It." *Forbes* 24 Mar. 1986: 56-66.

———. "You Mean We've Been Speaking Prose All These Years?" *Forbes* 11 Apr. 1983: 142-49.

Cooper, M., and M. Holzman. "Talking about Protocols." *College Composition and Communication* 34 (1983): 284-93.

Corder, Jim W. "On the Way, Perhaps, to a New Rhetoric, but Not There Yet, and If We Do Get There, There Won't Be There Anymore." *College English* 47 (1985): 161-70.

Couture, Barbara, et al. "Building a Professional Writing Program through a University/Industry Collaborative." Odell and Goswami, *Writing* 391-426.

Crane, Diane. *Invisible Colleges: Diffusion of Knowledge in Scientific Communities.* Chicago: U of Chicago P, 1972.

Crouch, William George, and Robert L. Zetler. *A Guide to Technical Writing.* New York: Ronald, 1948.

Crow, Wendell C. *Communication Graphics.* Englewood Cliffs: Prentice, 1986.

Cunningham, Donald H., and Gerald Cohen. *Creating Technical Manuals: A Step-by-Step Approach to Writing User-Friendly Instructions.* New York: McGraw, 1984.

Cunningham, Donald H., and Herman A. Estrin, eds. *The Teaching of Technical Writing.* Urbana: NCTE, 1975.

De Beaugrande, Robert. *Text Production: Toward a Science of Composition.* Norwood: Ablex, 1984.

Debs, Mary Beth. "Collaborative Writing: A Study of Technical Writing in the Computer Industry." Diss. Rensselaer Polytechnic Inst., 1986.

———. "Strategies for Successful Collaborative Writing." Conf. Record of the IEEE Professional Communication Soc. 13-14 Oct. 1982. New York: Inst. of Electrical and Electronics Engineers, 1982. 16-19.

DEC Professional. Spring House: Professional, 1983.

DeGeorge, James, Gary A. Olson, and Richard Ray. *Style and Readability in Technical Writing: A Sentence-Combining Approach.* New York: Random, 1984.

Demoney, Jerry, and Susan E. Meyer. *Pasteups and Mechanicals.* New York: Watson, 1982.

Denise, Richard M. "Technology for the Executive Thinker." *Datamation* June 1983: 206-16.

Dobrin, David N. "Protocols Once More." *College English* 48 (1986): 713-25.

———. "What's the Purpose of Teaching Technical Communication." *Technical Writing Teacher* 12 (1985): 146-60.

Doheny-Farina, Stephen. "Writing in an Emerging Organization." *Written Communication* 3.2 (1986): 158-85.

Doty, Kathleen. "Redefining the Writing Process: The Integration of Form and Content." Proc. of the Intl. Technical Communication Conf. 11-14 May 1986. Washington: Soc. for Technical Communication, 1986. 425-29.

Dressel, Susan. "Effective Collaboration for Authors and Technical Communicators." Conf. Record of the IEEE Professional Communication Soc. 10-12 Oct. 1984. New York: Inst. of Electrical and Electronics Engineers, 1984. 107-10.

Dreyfus, Hubert, and Stuart Dreyfus. "Why Computers May Never Think like People." *Technology Review* Jan. 1986: 42-61.

Drucker, Peter. "The Coming of the New Organization." *Harvard Business Review* Jan.-Feb. 1988: 45-53.

Duffy, T. M. "Readability Formulas: What's the Use?" *Designing Usable Texts.* Ed. T. M. Duffy and R. M. Waller. Orlando: Academic, 1985. 113-43.

Dunlap, Louise. "The 'Deskilling' of Writing in the Professional Workplace." Penn State Conf. on Rhetoric and Composition. University Park, 1985.

Earle, Samuel C. *The Theory and Practice of Technical Writing.* New York: Macmillan, 1911.

Ede, Lisa. "Audience Analysis: An Introduction to Research." *College Composition and Communication* 35 (1984): 140-55.

Ede, Lisa, and Andrea Lunsford. "Audience Addressed/Audience Invoked: The Role of Audience in Composition Theory and Pedagogy." *College Composition and Communication* 35 (1984): 155-71.

———. "Research into Collaborative Writing." *Technical Communication* 32.4 (1985): 69-70.

Ehrbar, A. F. "Grasping the New Unemployment." *Fortune* 16 May 1983: 106-12.

Ehrenhalt, Samuel M. "The Nature of Education—College Graduates and the Market." *Current* Nov. 1983: 15-24.

Emig, Janet. *The Composing Processes of Twelfth Graders.* NCTE Rept. 13. Urbana: NCTE, 1971.

Engel, Stephen E., and Richard E. Granda. *Guidelines for Man/Display Interfaces.* Technical Rept. 00.2720. Poughkeepsie: IBM, 1975.

Ericsson, K. A., and H. A. Simon. *Protocol Analysis: Verbal Reports as Data.* Cambridge: MIT P, 1985.

Estes, Susan Page, and Andrew Rojecki. "Publishing Secure Technical Literature with a Small Writing Staff." Proc. of the Intl. Technical Communication Conf. 19 Apr.–2 May 1984. Washington: Soc. for Technical Communication, 1984. ATA-99-102.

Faigley, Lester. "Competing Theories of Process." *College English* 48 (1986): 527-42.

Faigley, Lester, and Thomas Miller. "What We Learn from Writing on the Job." *College English* 44 (1982): 557-69.

Faigley, Lester, et al. *Assessing Writers' Knowledge and Processes of Composing.* Norwood: Ablex, 1985.

———. "Nonacademic Writing: The Social Perspective." Odell and Goswami, *Writing* 231-48.

———. *Writing after College: A Stratified Survey of the Writing of College-Trained People.* Technical Rept. 1. FIPSE Grant G008005896, 1981.

Feingold, S. Norman. "Emerging Careers: Occupations for Post-industrial Society." *Futurist* Feb. 1984: 9-16.

Felker, Daniel B., ed. *Document Design: A Review of Relevant Research.* Washington: American Insts. for Research, 1980.

Felker, Daniel B., et al. *Guidelines for Document Designers.* Washington: American Insts. for Research, 1981.

Finn, Seth. "Information-Theoretic Measures of Reader Enjoyment." *Written Communication* 2 (1985): 358-76.

Fisher, Harold A. "Broadcast Journalists' Perceptions of Appropriate Career Preparation." *Journalism Quarterly* 55 (1978): 140-44.

Fitting, Ralph U. *Report Writing.* New York: Ronald, 1924.

Flesch, Rudolf. *The Art of Readable Writing.* New York: Harper, 1974.

Flower, Linda. *Problem-Solving Strategies for Writing.* New York: Harcourt, 1981.

Flower, Linda, and John R. Hayes. "A Cognitive Process Theory of Writing." *College Composition and Communication* 32 (1981): 365-87.

———. "Identifying the Organization of Writing Processes" and "The Dynamics of Composing: Making Plans and Juggling Constraints." Gregg and Steinberg 3-50.

———. "Problem-Solving Strategies and the Writing Process." *College English* 39 (1977): 449-61.

Flower, Linda, John R. Hayes, and Heidi Swarts. "Revising Functional Documents: The Scenario Principle." Anderson, Brockmann, and Miller 41-58.

Fodor, J. A., and M. Garrett. "Some Syntactic Determinants of Sentential Complexity." *Perception and Psychophysics* 2 (1967): 289-96.

Foreman, Janis, and Patricia Katsky. "The Group Report: A Problem in Small Group or Writing Processes?" *Journal of Business Communication* 23.4 (1986): 23-36.

Fountain, Alvin M. *A Study of Courses in Technical Writing.* Diss. Peabody Coll., 1938.

Fraker, Susan. "High-Speed Management for the High-Tech Age." *Fortune* 5 Mar. 1984: 62-68.

"Frontinus' Classic Report." *Engineers as Writers.* Ed. W. J. Miller and L. E. A. Saidla. New York: Van Nostrand, 1953. 27-63.

Fry, Edward B. *Graphical Comprehension: How to Read and Make Graphs.* Providence: Jamestown, 1981.

Fry, Robert. "Expanding Concepts of the Writer's Purpose, Audience, and Task: *The IEEE Transactions on Professional Communication.*" *IEEE Transactions on Professional Communication* PC-30.1 (1987): 4-11.

Galbraith, John Kenneth. *The New Industrial State.* 2nd ed. New York: NAL, 1971.

Gallagher, William J. *Report Writing for Management.* Reading: Addison, 1969.

Garver, Eugene. "Teaching Writing and Teaching Virtue." *Journal of Technical Writing and Communication* 22.1 (1985): 51-73.

Gaum, Carl G., and Harold F. Graves. *Report Writing.* New York: Prentice, 1929.

Gaum, Carl G., Harold F. Graves, and Lynne S. S. Hoffman. *Report Writing.* 3rd ed. New York: Prentice, 1950.

Gebhardt, Richard. "Teamwork and Feedback: Broadening the Base of Collaborative Writing." *College English* 42 (1980): 69-74.

Gloe, Esther M. "Setting up Internships in Technical Writing." *Journal of Technical Writing and Communication* 13 (1983): 7-27.

Goble, Lynne. "The Author-Editor Interface." Conf. Record of the IEEE Professional Communication Soc. 10-12 Oct. 1984. New York: Inst. of Electrical and Electronics Engineers, 1984. 115-17.

Goswami, Dixie, et al. *Writing in the Professions.* Washington: American Insts. for Research, 1981.

Gough, Philip B. "Grammatical Transformations and Speed of Understanding." *Journal of Verbal Learning and Verbal Behavior* 4 (1965): 107-11.

Gould, John D. "Experiments on Composing Letters: Some Facts, Some Myths, and Some Observations." Gregg and Steinberg 118-26.

Grammatik III. Computer software. Reference Software, 1988.

Graves, Richard L. "Symmetrical Form and the Rhetoric of the Sentence." *Rhetoric and Composition: A Sourcebook for Teachers and Writers.* 2nd ed. Ed. Richard L. Graves. Upper Montclair: Boynton, 1984. 119-27.

Gray, Paul. "New Information Careers." *Business Week's Guide to Careers* Spring-Summer 1984: 69-74.

Grayson, C. Jackson, Jr. "Networking by Computer." *Futurist* June 1984: 14-17.

———. "Productivity On-line." *Across the Board* Jan. 1984: 30-35.

Green, Marcus, and Timothy D. Nolan. "A Systematic Analysis of the Technical Communicator's Job: A Guide for Educators." *Technical Communication* 31.4 (1984): 9-12.

Green, Petrina, and Ted Brooks. *How Technical Writers Write: A Survey of Strategies and Attitudes.* TR 21.1049. Kingston: IBM, 1987.

Greenbaum, Sidney. *Studies in Adverbial English Usage.* London: Longman, 1969.

Gregg, Lee W., and E. R. Steinberg, eds. *Cognitive Processes in Writing.* Hillsdale: Erlbaum, 1980.

Griesinger, Walter S., and Richard R. Klene. "Readability of Introductory Psychology Textbooks: Flesch versus Student Ratings." *Teaching of Psychology* 11 (1984): 90-91.

Griffiths, J. M. "Competency Requirements for Library and Information Science Professionals." *Information Science Abstracts* 20 (1985): 85-2017.

Grimm, Susan J. *How to Write Computer Manuals for Users*. Belmont: Wadsworth, 1982.

"The Growing Need for Education." (Projection 2000) *Occupational Outlook Quarterly* 31 (Fall 1987): 34-36.

Gunning, Robert. *The Technique of Clear Writing*. New York: McGraw, 1952.

Hagstrom, Warren O. "Traditional and Modern Forms of Scientific Teamwork." *Administrative Science Quarterly* 9 (1964): 241-63.

Hall, Dennis R. "The Role of Invention in Technical Writing." *Technical Writing Teacher* 4 (1976): 13-24.

Halpern, Jeanne W., and Sarah Liggett. *Computers and Composing*. Urbana: NCTE, 1984.

Harbarger, Sada A. *English for Engineers*. New York: McGraw, 1923.

Hardcopy. Placentina: Seldin.

Harrison, Teresa M. "Framework for the Study of Writing in Organizational Contexts." *Written Communication* 4.1 (1987): 3-23.

Hastings, G. Prentice, and Kathryn J. King. *Creating Effective Documentation for Computer Programs*. Englewood Cliffs: Prentice, 1986.

Hawkins, John A. *Definiteness and Indefiniteness: A Study in Reference and Grammaticality Prediction*. London: Croom, 1978.

Hays, Robert William. *Principles of Technical Writing*. Reading: Addison, 1965.

Helgeson, Donald V. *Writing Technical Proposals That Win Contracts*. Englewood Cliffs: Prentice, 1985.

Henderson, Allan. "The Care and Feeding of the Non-captive Reader." *Technical Communication* 31.3 (1984): 5-8.

Herrington, Anne. "Writing in Academic Settings: A Study of the Contexts for Writing in Two-Year-College Chemical Engineering Courses." *Research in the Teaching of English* 19 (1985): 331-61.

Hiatt, Mary P. *Artful Balance: The Parallel Structures of Style*. New York: Teachers Coll. P, 1975.

Hill, James, ed. *Management*. Spec. issue of *IEEE Transactions on Professional Communication* PC-28.8 (1985): 1-75.

Hine, Virginia H. "Networks in a Global Society." *Futurist* June 1984: 11-13.

Hird, Kenneth F. *Paste-up for Graphic Arts Production*. Englewood Cliffs: Prentice, 1982.

Hockley, H. R. Quoted by J. D. McIntosh. "Forum 75." *Communicator* 27.4 (1976): 4-5.

Holcombe, Marya W., and Judith K. Stein. *Writing for Decision Makers*. Belmont: Lifetime, 1981.

Hollister, Lahnice. "How to Listen to Technical Information." Conf. Record of the IEEE Professional Communication Soc. 16-18 Oct. 1985. New York: Inst. of Electrical and Electronics Engineers, 1985. 119-21.

Holtz, Herman, and Terry Schmidt. *The Winning Proposal: How to Write It*. New York: McGraw, 1981.

Horowitz, Irving. "Printed Words, Computers, and Democratic Societies." *Virginia Quarterly Review* 59 (1983): 620-36.

Horvath, Frances W. "The Pulse of Economic Change: Displaced Workers of 1981-1985." *Monthly Labor Review* June 1987: 3-13.

Houp, Kenneth W., and Thomas E. Pearsall. *Reporting Technical Information*. New York: Macmillan, 1968.

Howell, John Bruce. *Style Manuals of the English-Speaking World: A Guide.* Phoenix: Oryx, 1983.

Huddleston, Rodney D. *The Sentence in Written English: A Syntactic Study Based on an Analysis of Scientific Texts.* Cambridge: Cambridge UP, 1971.

Hull, Daniel M., and Leno S. Pedrotti. "Challenges and Changes in Engineering Technology." *Engineering Education* 76 (1986): 726-32.

International Data Corp. "Office Systems for the Eighties: Automation and the Bottom Line." *Fortune* 3 Oct. 1983: 89-162.

———. "The Revolution in Business Communication: How Businesses Can Cope." *Fortune* 19 Mar. 1984. 151-86.

Jackson, Tom. "Wake Up, Corporate America." *Business Week's Guide to Careers* Dec. 1984–Jan. 1985: 43-44.

Jenkins, Roger L., and Richard C. Reizenstein. "Insights into the MBA: Its Contents, Output, and Relevance." *Selections* (Graduate Management Admissions Council) 1 (1984): 19-24.

Johnson, Thomas P. *Analytical Writing: Handbook for Business and Technical Writing.* New York: Harper, 1966.

———. "How Well Do You Communicate?" *IEEE Transactions on Professional Communication* PC-25.2 (1982): 5-9.

Jones, Barry O. *Sleepers, Wake! Technology and the Future of Work.* New York: Oxford UP, 1982.

Jones, Walter Paul. *Writing Scientific Papers and Reports.* Dubuque: Brown, 1946.

Jordan, Michael. "Close Cohesion with *Do So*: A Linguistic Experiment in Language Function Using a Multi-example Corpus." *Functional Approaches to Writing: Research Perspectives.* Ed. Barbara Couture. London: Pinter, 1986. 29-48.

———. "The Principal Semantics of the Nominal 'This' and 'That' in Contemporary English." Diss. Hatfield Polytechnic, 1978.

———. "Some Clause-Relational Associated Nominals in Technical English." *Technostyle* 4.1 (1985): 36-46.

Jordan, Stello, ed. *Handbook of Technical Writing Practices.* 2 vols. New York: Wiley, 1971.

Journet, Debra, and Julie Lepick Kling, eds. *Readings for Technical Writers.* Dallas: Scott, 1984.

Just, Marcel Adam, and Patricia A. Carpenter. "A Theory of Reading: From Eye Fixations to Comprehension." *Psychological Review* 87 (1980): 329-54.

Kane, Thomas S. *The Oxford Guide to Writing: A Rhetoric and Handbook for College Students.* New York: Oxford UP, 1983.

Kapp, Reginald O. *The Presentation of Technical Information.* London: Constable, 1948. New York: Macmillan, 1957.

Keene, Michael L. *Effective Professional Writing.* Lexington: Heath, 1987.

Keene, Michael L., and Marilyn Barnes-Ostrander. "Audience Analysis and Adaptation." Moran and Journet 163-91.

Kelley, Patrick M., et al. *Academic Programs in Technical Communication.* 3rd ed. Washington: Soc. for Technical Communication, 1985.

Kendel, Stephen, and Ellen Benoir. "Hello, Mr. Chips." *Forbes* 23 Apr. 1984: 132.

Kennedy, George A. *Classical Rhetoric and Its Christian and Secular Tradition from Ancient to Modern Times.* Chapel Hill: U of North Carolina P, 1980.

Kerekes, Frank, and Robley Winfry. *Report Presentation.* Dubuque: Brown, 1948.

Kerr, Clark. *The Uses of the University.* New York: Harper, 1966.

Kiesler, Sarah, Jane Siegel, and Timothy W. McGuire. "Social Psychological Aspects of Computer-Mediated Communication." *American Psychologist* 39 (1984): 1123-34.

Kinneavy, James L. *A Theory of Discourse: The Aims of Discourse.* New York: Norton, 1971.

Kiplinger, Knight. "The Shape of Things to Come." *Changing Times* Jan. 1987: 28-47.

Klapp, Orin. "Meaning Lag in the Information Society." *Communication* Spring 1983: 37-53.

Knoblauch, C. H. "Intentionality in the Writing Process: A Case Study." *College Composition and Communication* 31 (1980): 153-59.

Kolers, Paul A. "Skill in Reading and Memory." *Canadian Journal of Psychology* 39 (1985): 232-39.

Kotter, John P. "What Effective General Managers Really Do." *Harvard Business Review* Nov.-Dec. 1982: 156-67.

Krohn, Roger G. "Patterns of the Institutionalization of Research." *The Social Contexts of Research.* Ed. Saad Z. Nagi and Ronald G. Corwin. New York: Wiley-Interscience, 1972. 29-66.

Kroll, Barry. "Writing for Readers: Three Perspectives on Audience." *College Composition and Communication* 35 (1984): 172-85.

Kuhn, Thomas. *The Structure of Scientific Revolutions.* 2nd ed. Chicago: U of Chicago P, 1970.

Kutscher, Ronald E. "An Overview of the Year 2000." *Occupational Outlook Quarterly* 31 (1988): 2-9.

Lampe, David R. "Engineer's Invisible Activity: Writing." *Technology Review* Apr. 1983: 73-74.

Landvater, Darryl. "Technology—A Catalyst for Human Change." *Infosystems* May 1984: 84.

Lannon, John M. *Technical Writing.* 3rd ed. Boston: Little, 1985.

Larson, Virginia. *How to Write a Winning Proposal.* 2nd ed. San Diego: Classic, 1986.

Lauer, David A. *Design Basics.* 2nd ed. New York: Holt, 1985.

Leech, Geoffrey N. *A Linguistic Guide to English Poetry.* London: Longman, 1969.

Lefferts, Robert. *How to Prepare Charts and Graphs for Effective Reports.* New York: Barnes, 1981.

Leonard, David C., and Peter J. McGuire, eds. *Readings in Technical Communication.* New York: Macmillan, 1983.

Lindemann, Erika. *Longman Bibliography of Composition and Rhetoric: 1984–1985.* New York: Longman, 1987.

———. *Longman Bibliography of Composition and Rhetoric: 1986.* New York: Longman, 1988.

Lipson, Marjorie Y. "Learning New Information from Text: The Role of Prior Knowledge and Reading Ability." *Journal of Reading Behavior* 14 (1982): 243-61.

Lobkowicz, Nicholas. *Theory and Practice: History of a Concept from Aristotle to Marx.* Notre Dame: U of Notre Dame P, 1967.

Lodahl, Janice B., and Gerald Gordon. "The Structure of Scientific Fields and the Functioning of University Graduate Departments." *American Sociological Review* 37 (1972): 57-72.

Long, Russell. "Writer-Audience Relationships: Analysis or Invention?" *College Composition and Communication* 31 (1980): 221-26.

Longacre, Robert E. *Hierarchy and Universality of Discourse Constituents in New Guinea Languages.* 2 vols. Washington: Georgetown UP, 1972.

Lotus 1-2-3. Version 2.0. Computer software. Lotus, 1986.

Lunsford, Andrea, and Lisa Ede. "Why Write . . . Together: A Research Update." *Rhetoric Review* 5.1 (1986): 71-81.

Lutz, Jean. "Writers in Organizations and How They Learn the Image: Theory, Research, and Implications." *Worlds of Writing.* Ed. Carolyn Matalene. New York: Random, forthcoming.

Lynch, Robert E., and Thomas B. Swanzey, eds. *The Example of Science: An Anthology for College Composition.* Englewood Cliffs: Prentice, 1981.

MacGregor, A. J. *Graphics Simplified: How to Plan and Prepare Effective Charts, Graphs, Illustrations, and Other Visuals.* Toronto: U of Toronto P, 1979.

MacIntyre, Alasdair. *After Virtue: A Study in Moral Theory.* 2nd ed. Notre Dame: U of Notre Dame P, 1984.

Magee, John, and Arthur D. Little, Inc. "SMR Forum: What Information Technology Has in Store for Managers." *Sloan Management Review* 26.2 (1985): 45-49.

Maisonrouge, Jacques G. "The Education of a Modern International Manager." *Journal of International Business Studies* 14 (1983): 141-46.

Mann, William C., and Sandra A. Thompson. "Relational Propositions in Discourse." Rept. ISI/RR-83-115. Marina del Rey: Instructional Sciences Inst., U of Southern California, 1973.

———. "Rhetorical Structure Theory: A Theory of Text-Organization." Rept. ISI/RS-87-190. Marina del Rey: Instructional Sciences Inst., U of Southern California, 1987.

Marder, Daniel. *The Craft of Technical Writing.* New York: Macmillan, 1960.

Marra, James L. "For Writers: Understanding the Art of Layout." *Technical Communication* 28.3 (1981): 11-13, 40.

Mathes, J. C. "Prospects for Technical Communication: Research for Future Needs." Conf. on College Composition and Communication. Minneapolis, 1979.

Mathes, J. C., and Dwight W. Stevenson. *Designing Technical Reports: Writing for Audiences in Organizations.* Indianapolis: Bobbs, 1976.

Mattill, John I. "High Stakes in Lifelong Education." *Technology Review* Jan. 1983: 81, 86.

McClure, Glenda M. "Readability Formulas: Useful or Useless?" *IEEE Transactions on Professional Communication* PC-30.1 (1987): 12-15.

McInnis, Noel. "Networking: A Way to Manage Our Changing World?" *Futurist* June 1984: 9-10.

McKeon, Richard. "Aristotle's Conception of Language and the Arts of Language." *Critics and Criticism, Ancient and Modern.* Ed. R. S. Crane. Chicago: U of Chicago P, 1952. 176-231.

McKinley, Lawrence C. "Managing for User-Friendly Publications." Proc. of the Second Intl. Conf. on Systems Documentation. 28-30 Apr. 1983. New York: Assoc. for Computer Machinery, 1983. 74-81.

McKoon, Gail, et al. "A Critical Evaluation of the Semantic-Episodic Distinction." *Journal of Experimental Psychology: Learning, Memory, and Cognition* 12 (1986): 295-306.

McVey, Gerald. "Legibility in Film-Based and Television Display Systems." *Technical Communication* 32.4 (1985): 21-28.

Menofsky, Joseph A. "Computer Workship: Faster Learning, Higher Grades, Better Jobs. Computer Literacy Promises All That, and Heaven Too." *Science 84* May 1984: 40-46.

Merton, Robert K. "The Behavior Patterns of Scientists." *American Scientist* 57 (1969): 1-23.

———. *The Sociology of Science: Theoretical and Empirical Investigations.* Chicago: U of Chicago P, 1973.

Meyer, B. D. "The ABCs of New-Look Publications." *Technical Communication* 33 (1986): 16-20.

Miles, Josephine. *Style and Proportion: The Language of Prose and Poetry.* Boston: Little, 1967.

Miller, Carolyn R. "Environmental Impact Statements and Rhetorical Genres: An Application of Rhetorical Theory to Technical Communication." Diss. Rennselaer Polytechnic Inst., 1980.

———. "Genre as Social Action." *Quarterly Journal of Speech* 70 (1984): 151-67.

———. "A Humanistic Rationale for Technical Writing." *College English* 40 (1979): 610-17.

———. "Invention in Technical and Scientific Discourse: A Prospective Survey." Moran and Journet 116-62.

———. "Technology as a Form of Consciousness: A Study of Contemporary Ethos." *Central States Speech Journal* 29 (1978): 228-36.

Miller, Carolyn R., and Jack Selzer. "Special Topics of Argument in Engineering Reports." Odell and Goswami, *Writing* 309-41.

Mills, C. B., and K. L. Dye. "Usability Testing: User Reviews." *Technical Communication* 32.4 (1985): 40-44.

Mills, Gordon H., and John A. Walter. *Technical Writing.* New York: Holt, 1954.

"Mission Statement." North Carolina State University Board of Trustees. 1984.

Mitchell, John H. "British Techniques for Teaching Technical Writing." *Society of Technical Writers and Publishers Review* 10.2 (1962): 19-22.

———. *Handbook of Technical Communication.* Belmont: Wadsworth, 1962.

———. "Teaching Technical Writing in Other Countries." *Journal of Technical Writing and Communication* 8 (1978): 113-19.

Mitchell, Ruth, and Mary Taylor. "The Integrating Perspective: An Audience-Response Model for Writing." *College English* 41 (1979): 247-71.

Moore, Steve. "Information Managers Must Face the International Communication Web." *Data Management* June 1984: 30-32.

Moran, Michael G., and Debra Journet, eds. *Research in Technical Communication: A Bibliographic Sourcebook.* Westport: Greenwood, 1985.

Moriarty, Sandra E., and Edward C. Scheiner. "A Study of Close-Set Text Type." *Journal of Applied Psychology* 69 (1984): 700-02.

Morton, John, et al. "Headed Records: A Model for Memory and Its Failures." *Cognition* 20 (1985): 1-23.

Mosenthal, Peter B., et al. "The Influence of Prior Knowledge and Teacher Lesson Structure on Children's Productions of Narratives." *Elementary School Journal* 85 (1985): 621-34.

Movshon, J. Anthony. *FPRINT Font Printer.* Version 2.0. Computer software, 1985. [100 Bleecker St., New York, NY 10012]

Murch, Gerald M. "Using Color Effectively: Designing to Human Specifications." *Technical Communication* 32.4 (1985): 14-20.

Murray, Donald. "Internal Revision: A Process of Discovery." *Research on Composing: Points of Departure.* Ed. Charles Cooper and Lee Odell. Urbana: NCTE, 1978. 85-103.

———. "Write before Writing." *College Composition and Communication* 29 (1978): 375-81.

———. "Writing as Process: How Writing Finds Its Own Meaning." *Eight Approaches to Teaching Composition.* Ed. Timothy R. Donovan and Ben W. McClelland. Urbana: NCTE, 1980. 3-20.

Myers, Greg. "Texts as Knowledge Claims: The Social Construction of Two Biologists' Articles." *Social Studies of Science* 15 (1985): 593-630.

Nash, Walter. *Designs in Prose.* London: Longman, 1980.

Neiburger, Carl D. *TypeStyl.* Version 1.1. Computer software, 1987. [169 N. 25 St., San Jose, CA 95116]

Nelms, Henning. *Thinking with a Pencil.* Berkeley: Ten Speed, 1981.

Nelson, Joseph Raleigh. *Writing the Technical Report.* New York: McGraw, 1940.

Nelson, Milo. "Information vs. Knowledge." *Wilson Library Bulletin* Mar. 1980: 420-22.

Nelson, Roy Paul. *The Design of Advertising.* 5th ed. Dubuque: Brown, 1985.

Nesbit, R. E., and T. D. Wilson. "Telling More than We Can Know: Verbal Reports on Mental Processes." *Psychological Review* 84 (1977): 231-59.

Ng, Seok Moi. "A Method for Assessing the Readability of Reading Texts in the Junior School." *New Zealand Journal of Educational Studies* 19 (1984): 76-79.

Odell, Lee. "Beyond the Text: Relations between Writing and Social Context." Odell and Goswami, *Writing* 249-80.

Odell, Lee, and Dixie Goswami. "Writing in a Nonacademic Setting." *New Directions in Composition Research.* Ed. Richard Beach and Lillian S. Bridwell. New York: Guilford, 1984. 233-58.

———, eds. *Writing in Nonacademic Settings.* New York: Guilford, 1985.

Odell, Lee, Dixie Goswami, and Anne Herrington. "The Discourse-Based Interview: A Procedure for Exploring Tacit Knowledge of Writers in Non-academic Settings." *Research on Writing: Principles and Methods.* Ed. Peter Mosenthal et al. New York: Longman, 1983. 221-36.

———. "Studying Writing in Non-academic Settings." Anderson, Brockmann, and Miller 17-40.

Ohmann, Richard. *English in America: A Radical View of the Profession.* New York: Oxford UP, 1976.

Olsen, Leslie A., and Thomas N. Huckin. *Principles of Communication for Science and Technology.* New York: McGraw, 1983.

Ong, Walter J. "The Writer's Audience Is Always a Fiction." *PMLA* 90 (1975): 9-21.

Pakin, Sandra. *Documentation Development Methodology: Techniques for Improved Communications.* Englewood Cliffs: Prentice, 1982.

Palokoff, Kathy. "How Management Decisions Affect Writer-Artist Collaboration." Proc. of the 32nd Intl. Technical Communication Conf. 19-22 May 1985. Washington: Soc. for Technical Communication, 1985. MPD32-34.

Paradis, James, David Dobrin, and Richard Miller. "Writing at Exxon ITD: Notes on the Writing Environment of an R&D Organization." Odell and Goswami, *Writing* 281-309.

Park, Douglas B. "Analyzing Audiences." *College Composition and Communication* 37 (1986): 478-88.

————. "The Meaning of Audience." *College English* 44 (1982): 247-57.

Parret, Tom. "Mendacious Machines." *Datamation* Apr. 1984: 134-37.

Pearsall, Thomas E. *Audience Analysis for Technical Writing.* Beverly Hills: Glencoe, 1969.

————. *Teaching Technical Writing: Methods for College English Teachers.* Washington: Soc. for Technical Communication, 1975.

Pearsall, Thomas E., and Frances Sullivan. *Academic Programs in Technical Communication.* Washington: Soc. for Technical Communication, 1976.

Pearsall, Thomas E., Frances Sullivan, and Earl McDowell. *Academic Programs in Technical Communication.* 2nd ed. Washington: Soc. for Technical Communication, 1981.

Perrin, Porter G. *Writer's Guide and Index to English.* Chicago: Scott, 1942.

Peters, Thomas J., and Nancy Austin. *A Passion for Excellence: The Leadership Difference.* New York: Random, 1985.

Peters, Thomas J., and Robert H. Waterman. *In Search of Excellence: Lessons from America's Best-Run Companies.* New York: Harper, 1982.

Peuto, Bernard L. "What Writers Can Expect from an Electronic Publishing System." Proc. of the Intl. Technical Communication Conf. 19 Apr.–2 May 1984. Washington: Soc. for Technical Communication, 1984, ATA-11-14.

Pfister, Fred R., and Joanne F. Petrick. "A Heuristic Model for Creating a Writer's Audience." *College Composition and Communication* 31 (1980): 213-20.

PFS: Graph. Computer software. Software Publishing Corp., 1984.

Pinelli, Thomas E., Virginia M. Cordle, and Robert McCullough. "A Survey of Typography, Graphic Design, and Physical Media in Technical Reports." *Technical Communications* 33 (1986): 75-80.

Pinelli, Thomas E., et al. "Report Format Preferences of Technical Managers and Nonmanagers." *Technical Communication* 31.2 (1984): 4-8.

Pittel, Leslie. "Existential Accounting." *Forbes* 30 Aug. 1984: 104-08.

Poter-Roth, Bud. *Proposal Development: A Winning Approach.* Milpitas: PSI Research, 1984.

Price, Derek J. de Solla, and Donald Beaver. "Collaboration in an Invisible College." *American Psychologist* 21 (1966): 1011-18.

Publication Manual of the American Psychological Association. 3rd ed. Washington: American Psychological Assn., 1983.

Publish! San Francisco: PCW Communications.

Quirk, Randolph, et al. *A Grammar of Contemporary English.* London: Longman, 1972.

Ramey, Judy. "Developing a Theoretical Base for On-line Documentation. Part I: Building a Theory." *Technical Writing Teacher* 13 (1986): 148-59.

————. "Developing a Theoretical Base for On-line Documentation. Part II: Applying the Theory." *Technical Writing Teacher* 13 (1986): 302-15.

Ratcliff, Roger, and Gail McKoon. "More on the Distinction between Episodic and Semantic Memories." *Journal of Experimental Psychology: Learning, Memory, and Cognition* 12 (1986): 312-13.

Rathbone, Robert R., and James B. Stone. *A Writer's Guide for Engineers and Scientists.* Englewood Cliffs: Prentice, 1962.

Ravitch, Diane. "The Educational Pendulum." *Psychology Today* Oct. 1983: 62-71.

Readability Calculations according to Nine Formulas. Computer software. Micro Power and Light, 1984.

Redish Janice C. "The Language of the Bureaucracy." *Literacy for Life: The Demand for Reading and Writing.* Ed. Richard W. Bailey and Robin Melanie Fosheim. New York: MLA, 1983. 151-74.

———."Writing in Organizations." *Writing in the Business Professions.* Ed. Myra Kogen. Urbana: NCTE, 1989. 97-124.

Redish, Janice C., R. M. Battison, and E. S. Gold. "Making Information Accessible to Readers." Odell and Goswami, *Writing* 129-53.

Reed, Jeffrey G. "Preparing the Training and Development Specialist: Skills and Knowledge Essential for Practice." *Journal of Business Education* 60.3 (1984): 8-13.

Rehder, Robert R., and James L. Porter. "The Creative MBA: A New Proposal for Balancing the Science and the Art of Management." *Business Horizons* Nov.-Dec. 1983: 52-54.

Reich, Robert B. "The Next American Frontier." *Atlantic* Mar. 1983: 43-58.

Reither, James. "Writing and Knowing: Toward Redefining the Writing Process." *College English* 47 (1985): 620-28.

Rickard, T. A. *A Guide to Technical Writing.* San Francisco: Mining and Scientific, 1908.

———.*Technical Writing.* New York: Wiley, 1920.

Ridgway, Lenore S. "Information on Display Screens." Proc. of the Intl. Technical Communication Conf. 1-4 May 1983. Washington: Soc. for Technical Communication, 1983. ATA-18-21.

Roberts, David D., and Patricia A. Sullivan. "Beyond the Static Audience Construct: Reading Protocols in the Technical Writing Class." *Journal of Technical Writing and Communication* 14 (1984): 143-53.

Rose, Lisle A., Burney B. Bennett, and Elmer Heater. *Engineering Reports.* New York: Harper, 1950.

Roth, Robert G. "The Evolving Audience: Alternatives to Audience Accommodation." *College Composition and Communication* 38 (1987): 47-55.

Roundy, Nancy, and David Mair. *Strategies for Technical Communication.* Boston: Little, 1985.

Rubens, Philip. "A Reader's View of Text and Graphics: Implications for Transactional Text." *Journal of Technical Writing and Communication* 16 (1986): 73-86.

Rubens, Philip, and Robert Krull. "Application of Research on Document Design to Online Display." *Technical Communication* 32.4 (1985): 29-43.

Rubin, Donald L. "Social Cognition and Written Communication." *Written Communication* 1 (1984): 211-45.

Rudolph, Frederick. *Curriculum: A History of the American Undergraduate Course of Study since 1636.* San Francisco: Jossey, 1977.

Saar, Doreen Alvarez. "A Technical Writing Teacher Becomes a Technical Writer." Conf. Record of the IEEE Professional Communication Soc. 22-24 Oct. 1986. New York: Inst. of Electrical and Electronics Engineers, 1986. 131-34.

Sadowski, Mary. "Elements of Composition." *Technical Communication* 34 (1987): 29-30.

Salino, L. E. "Continuing Engineering Education: One Element of Lifelong Learning for Engineers." *IEEE Transactions on Education* E 26.4 (1983): 122-26.

Salvatori, Mariolina. "Reading and Writing a Text: Correlations between Reading and Writing." *College English* 45 (1983): 657-66.

Sanderlin, Reed. "Information Is Not Communication." *Business Horizons* Mar.-Apr. 1982: 40-42.

Savin, Harris, and Ellen Perchonock. "Grammatical Structure and the Immediate Recall of English Sentences." *Journal of Verbal Learning and Verbal Behavior* 4 (1965): 348-53.

Sawyer, Thomas M., ed. *Technical and Professional Communication: Teaching in the Two-Year College, Four-Year College, Professional School.* Ann Arbor: Professional, 1977.

Schatzberg, Walter, Ronald A. Waite, and Jonathan K. Johnson, eds. *The Relations of Literature and Science: An Annotated Bibliography of Scholarship, 1880–1980.* New York: MLA, 1987.

Schell, D. A. "Testing Online User Documentation." *IEEE Transactions on Professional Communication* PC-29.4 (1986): 87-92.

———. "Usability Testing of Screen Design: Beyond Standards, Principles, and Guidelines." Proc. of the Human Factors Soc. Dayton, Oct. 1986. 1212-15.

Schiff, Jeff. "Who's Reading Whom?: An Audience Analysis Primer." *Arizona English Bulletin* 27 (1985): 19-24.

Schlefer, Jonathan. "Workers and Automation." *Technology Review* Feb.-Mar. 1983: 83.

Schoff, Gretchen H., and Patricia A. Robinson. *Writing and Designing Operator Manuals: Including Service Manuals and Manuals for International Markets.* Belmont: Lifetime, 1984.

Schön, Donald. *The Reflective Practitioner: How Professionals Think in Action.* New York: Basic, 1983.

Script. Computer software. IBM, n.d.

Seidel, Robert H., and Charles W. Gainey. *Script/PC.* Computer software. IBM, n.d.

Seifert, Colleen M., et al. "Types of Inferences Generated during Reading." *Journal of Memory and Language* 24 (1985): 405-22.

Selzer, Jack. "The Composing Processes of an Engineer." *College Composition and Communication* 34 (1983): 178-87.

———. "Exploring Options in Composing." *College Composition and Communication* 35 (1984): 276-84.

Sherman, Theodore A. *Modern Technical Writing.* New York: Prentice, 1955.

Sherman, Theodore A., and Simon S. Johnson. *Modern Technical Writing.* 4th ed. Englewood Cliffs: Prentice, 1983.

Sides, Charles H. *How to Write Papers and Reports about Computer Technology.* Philadelphia: ISI, 1984.

Silvers, Penny. "Process Writing and the Reading Connection." *Reading Teacher* 39 (1986): 684-88.

Skees, William D. *Writing Handbook for Computer Professionals.* Belmont: Lifetime, 1982.

Skillin, Marjorie, et al. *Words into Type.* 3rd ed. Englewood Cliffs: Prentice, 1974.

Smith, Anthony E. "The Information Revolution of the 1990s." *Political Quarterly* 54 (1983): 187-91.

Smith, Terry C. *How to Write Better and Faster.* New York: Crowell, 1965.

Snow, C. P. *The Two Cultures and the Scientific Revolution.* New York: Cambridge UP, 1959.

Society for Technical Communication. *Strategic Plan, 1986-1990.* [Washington: Soc. for Technical Communication, 1986.]

———. Ad Hoc Committee on Certification. "Certification of Technical Communicators." *Technical Communication* 27.1 (1980): 4-6, 15.

Soderston, Candace. "An Evaluative and Prescriptive Look at Graphics Research." Proc. of the Intl. Technical Communication Conf. 1-4 May 1983. Washington: Soc. for Technical Communication, 1983. RET-87-90.

———. "The Usability Edit: A New Level." *Technical Communication* 32.1 (1985): 16-18.

Sommers, Nancy. "Revision Strategies of Student Writers and Experienced Adult Writers." *College Composition and Communication* 31 (1980): 378-88.

Sopher, H. "Parallelism in Modern English Prose: Its Formal Patterns, Rhetorical Patterns, and Notional Relations." *English Studies* 63.1 (1982): 37-48.

Souther, James W. "Identifying the Informational Needs of Readers." *IEEE Transactions on Professional Communication* PC-28.8 (1985): 9-12.

———. *Technical Report Writing*. New York: Wiley, 1957.

———. "What Management Wants in the Technical Report." *Journal of Engineering Education* 52 (1962): 498-503.

Souther, James W., and Myron L. White. *Technical Report Writing*. 2nd ed. New York: Wiley, 1977.

Sparrow, W. Keats, and Nell Ann Pickett, eds. *Technical and Business Communication in Two-Year Programs*. Urbana: NCTE, 1983.

Sproull, Lee, and Sara Kiesler. "Reducing Social Context Cues: Electronic Mail in Organizational Communication." *Management Science* 32 (1986): 1492-512.

Steele, Lowell W. "Meeting International Competition." *Research Management* Mar.-Apr. 1984: 36-41.

Steinberg, Erwin R. "Protocols, Retrospective Reports, and the Stream of Consciousness." *College English* 48 (1986): 697-712.

Stevenson, Dwight W., ed. *Courses, Components, and Exercises in Technical Communication*. Urbana: NCTE, 1981.

———. "Toward a Rhetoric of Scientific and Technical Discourse." *Technical Writing Teacher* 5 (1976): 4-10.

Sticht, Thomas, et al. *The Role of Reading in the Navy*. NPRDC TR 77-40. San Diego: Navy Personnel Research and Development Center, 1977.

Stratton, Charles R. *Technical Writing: Process and Product*. New York: Holt, 1984.

Strunk, William, Jr., and E. B. White. *Elements of Style*. 3rd ed. New York: Macmillan, 1979.

Sullivan, Patricia, and Linda Flower. "How Do Users Read Computer Manuals? Some Protocol Contributions to Writers' Knowledge." *Convergences: Transactions in Reading and Writing*. Ed. Bruce T. Petersen. Urbana: NCTE, 1986. 163-78.

Swarts, Heidi, Linda Flower, and John R. Hayes. *How Headings in Documents Can Mislead Readers*. Document Design Project Rept. 9. Pittsburgh: Carnegie-Mellon U, 1980. ERIC ED 192344.

Sypherd, W. O., and Sharon Brown. *The Engineer's Manual of English*. New York: Scott, 1933.

Sypherd, W. O., Sharon Brown, and Alvin M. Fountain. *The Engineer's Manual of English*. Rev. ed. Chicago: Scott, 1943.

Sypherd, W. O., et al. *Manual of Technical Writing*. Chicago: Scott, 1957.

Tebeaux, Elizabeth. "Redesigning Professional Writing Courses to Meet the Communication Needs of Writers in Business and Industry." *College Composition and Communication* 36 (1985): 419-28.

"Technical Fields Hiring More Women to Rewrite Jargon." *Washington Post* 12 Mar. 1986: 31.

Thing, Lowell. "What the Well-Dressed Manual Is Wearing Today." *Technical Communication* 31.3 (1984): 8-12.

Thomas, Gordon P. "Mutual Knowledge: A Theoretical Basis for Analyzing Audience." *College English* 48 (1986): 580-94.

Thomas, Michael M. "Business Education: A Study in Paradox." *Business and Society* 27.1 (1983): 18-21.

Tichy, Henrietta J. *Effective Writing for Engineers, Managers, Scientists.* New York: Wiley, 1966.

Trelease, Sam Farlow, and Emma S. Yule. *Preparation of Scientific and Technical Papers.* Baltimore: Williams, 1925.

Trimble, John R. *Writing with Style: Conversations on the Art of Writing.* Englewood Cliffs: Prentice, 1975.

Tulving, Endel. "What Kind of Hypothesis Is the Distinction between Episodic and Semantic Memory?" *Journal of Experimental Psychology: Learning, Memory, and Cognition* 12 (1986): 307-11.

"The Two-Hatted M.B.A." *Forbes* 14 July 1986: 8-10.

Ulman, Joseph N., Jr. *Technical Reporting.* New York: Holt, 1952.

Ulman, Joseph N., Jr., and Jay R. Gould. *Technical Reporting.* 2nd ed. New York: Holt, 1959.

Underwood, N. R., and G. W. McConkie. "Perceptual Span for Letter Distinctions during Reading." *Reading Research Quarterly* 20 (1985): 153-62.

United States Army Material Development and Readiness Command. *An Introduction to the Preparation of New Look Technical Manuals.* Darcom Handbook, 310-1.1-80. Washington: GPO, 1981.

United States Congress. Office of Technology Assessment. "The Changing Nature of Office Work." *Automation of America's Offices, 1985–2000.* Washington: GPO, 1985. 95-104.

UNIX Review. San Francisco: Miller Freeman.

Veysey, Laurence R. *The Emergence of the American University.* Chicago: U of Chicago P, 1963.

Vreeland, James J. "Get It Wrong the First Time—So That You Can Get It Right by First Draft." Proc. of the USER-bility Annual Symposium. 31 July–1 Aug. 1984. Kingston: IBM, 1984. 5-8.

Walt, Harold R. "Toward 2000." *Management World* Jan. 1983: 8-10.

Warnock, John. "The Writing Process." *Research in Composition and Rhetoric: A Bibliographic Sourcebook.* Ed. Michael Moran and Ronald Lunsford. Westport: Greenwood, 1984. 3-26.

Warren, Thomas L. *Teachers Resources for Technical Writing: Purpose, Process, and Form.* Belmont: Wadsworth, 1984.

———.*Technical Writing: Purpose, Process, and Form.* Belmont: Wadsworth, 1984.

Watt, Homer A. *The Composition of Technical Papers.* New York: McGraw, 1917.

Weathers, Winston. "The Rhetoric of the Series." *Rhetoric and Composition: A Sourcebook for Teachers.* Ed. Richard L. Graves. Rochelle Park: Hayden, 1976. 95-101.

Weil, Benjamin Henry, et al., eds. *Technical Editing.* New York: Reinhold, 1958.

Weinberg, Alvin M. "Scientific Teams and Scientific Laboratories." *Daedalus* 99 (1970): 1056-75.

Weisman, Herman M. *Basic Technical Writing.* Columbus: Merrill, 1962.

———. *Technical Correspondence.* New York: Wiley, 1968.

Wells, Susan. "Jurgen Habermas, Communicative Competence, and the Teaching of Technical Discourse." *Theory in the Classroom.* Ed. Cary Nelson. Champaign: U of Illinois P, 1986. 245-69.

Wessel, David. "Computer Software for Writers: Helping the Bad, Hurting the Good?" *Wall Street Journal* 7 July 1986, eastern ed., sec. 2: 13.

Westwood, Albert R. C. "R&D Linkages in a Multi-Industry Corporation." *Research Management* May-June 1984: 23-26.

Whalen, Tim. *Writing and Managing Winning Technical Proposals.* Norwood: Artec, 1987.

White, Martha C. "The 1988–1989 Job Outlook in Brief." *Occupational Outlook Quarterly* 32 (1988): 10-16.

Whitehead, Alfred N. *The Aims of Education.* New York: Macmillan, 1929.

Willey, R. J., et al. "Teaching Audience Awareness to Freshman Writers." *Teaching English in the Two-Year College* 13 (1986): 7-11.

Winkler, Victoria M. "The Role of Models in Technical and Scientific Writing." Anderson, Brockmann, and Miller 111-22.

Winter, Eugene O. "A Clause-Relational Approach to English Texts: A Study of Some Predictive Lexical Items in Written Discourse." *Instructional Science* 6 (1977): 1-92.

———. "Replacement as a Function of Repetition: A Study of Some of Its Predictive Features in the Clause Relations of Contemporary English." Diss. U of London, 1974.

Wolff, Michael F. "What You Should Know about Teleconferencing." *Research Management* May-June 1984: 8-10.

———. "When Lab and Field Scientists Don't Communicate." *Research Management* Mar.-Apr. 1984: 7-8.

Writer's Workbench. Computer software. Bell, n.d.

Zuboff, Shoshana. "New Worlds of Computer-Mediated Work." *Harvard Business Review* Sept.-Oct. 1982: 145-52.

Zuckerman, Harriet. "Nobel Laureates in Science: Patterns of Productivity, Collaboration, and Authorship." *American Sociological Review* 32 (1967): 391-403.

Index

DEC 13 1991

LIBRARY OF MOUNT ST. MARYS
WITHDRAWN